Midédji Didier HOUELOKOU

Benin's security arrangements in the Gulf of Guinea

Midédji Didier HOUELOKOU

Benin's security arrangements in the Gulf of Guinea

ScienciaScripts

Imprint

Cover image: www.ingimage.com

This book is a translation from the original published under ISBN 978-620-2-27164-6.

Publisher:
Sciencia Scripts
is a trademark of
Dodo Books Indian Ocean Ltd. and OmniScriptum S.R.L publishing group

120 High Road, East Finchley, London, N2 9ED, United Kingdom
Str. Armeneasca 28/1, office 1, Chisinau MD-2012, Republic of Moldova, Europe
Managing Directors: Ieva Konstantinova, Victoria Ursu
info@omniscriptum.com

Printed at: see last page
ISBN: 978-620-8-51242-2

DEDICATION

This memoir is dedicated to :

- my wife and children. You are definitely the driving force behind my progress.
- I'm grateful to my parents, who did the essential work of enrolling me in school.

ACKNOWLEDGEMENTS

We would like to express our sincere thanks to all those who contributed to the production of this dissertation. In particular, we would like to thank :

Mr Léandre Edgard NDJAMBOU, CAMES Senior Lecturer, Teacher-Researcher, Omar Bongo University, whose invaluable assistance was decisive in carrying out this study, which he agreed to direct.

Mr Eugène Ogandaga of Omar Bongo University, whose guidance and availability were instrumental in the completion of this work.

To the management of the École d'État-major de Libreville (EEML) for their ongoing efforts to foster a culture of excellence. The daring reform of schooling at the EEML has made it possible to include this master's degree in the training. Our special thanks go to the Commandant of the school, Brigadier General MOUISSI Pamphile, the Director of Studies, Lieutenant-Colonel LESTIEN, and the Academic Advisor, Lieutenant-Colonel Arnaud Jézéquiel.

To our teachers at the Université Omar Bongo in Libreville, particularly those in the Geography Department, we pledge to keep alive the flame that you have rekindled in us during all these seminars.

Admiral Maxime Fernand AHOYO, Maritime Prefect of Benin, whose help and advice were invaluable in the preparation of this work.

To our group teachers, thank you for demanding the best from us.

All the officers of the French Navy who gave us some of their time to guide and advise us. Especially Lieutenant Major N'DAH Joas, for his availability.

My family, for their unwavering support.

My sincere thanks to all of you.

CONTENTS

GENERAL INTRODUCTION

Background and justification for the subject

The Gulf of Guinea is an area located in the west of the African continent with over 7,000 kilometres of coastline stretching from the north of Senegal to the south of Angola. The region comprises 16 countries, including Guinea-Bissau, Guinea-Conakry, Sierra Leone, Liberia, Côte d'Ivoire, Ghana, Togo, Benin, Nigeria, Cameroon, Equatorial Guinea, Sao Tome and Principe, Gabon, Congo, DRC and Angola. It is of strategic geopolitical importance due to its abundant natural resources such as oil, natural gas, minerals and fisheries. In addition, its position as a major transit route for international trade further reinforces its regional and global importance.

Benin, like other countries in the Gulf of Guinea, faces a number of major security challenges, from maritime piracy and drug trafficking to illicit arms trafficking. In a regional geopolitical context where the assertion of sovereignty is increasingly taking belligerent forms, analysing Benin's security arrangements in the Gulf of Guinea with a view to identifying shortcomings and recommending solutions is part of the anticipatory approach that should characterise any State that is aware of its strategic assets. Given that the emergence of the jihadist terrorist threat in the northern region has become a priority for both the government and Benin's defence and security forces since 2021, and that maritime security issues are likely to be relegated to second place, it is of major interest to reflect on the national security system in the Gulf of Guinea, so that a state of watch can be maintained on this issue.

As part of a collective security approach, the country participates in regional and international initiatives aimed at combating these threats, notably through regional cooperation initiatives (Yaoundé Architecture, Gulf of Guinea Commission, etc.) and international initiatives (France, USA, China, etc.). For a country with a coastline on the Atlantic Ocean, safety at sea should not remain a preoccupation for maritime specialists alone. It needs to be decompartmentalised to facilitate the emergence of a maritime culture.[1]

[1] In the sense of "maritime awareness" defined by the Lomé Charter as the effective understanding of everything relating to the maritime domain that could have an impact on safety, security, the economy or the environment, maritime culture for us extends beyond these considerations. For us, maritime culture extends beyond these considerations. It is also education about the sea and its related activities, so that it becomes a reflex that is systematically taken into account in all public policies relating to the sea.

Map No.1 : Countries in the Gulf of Guinea.

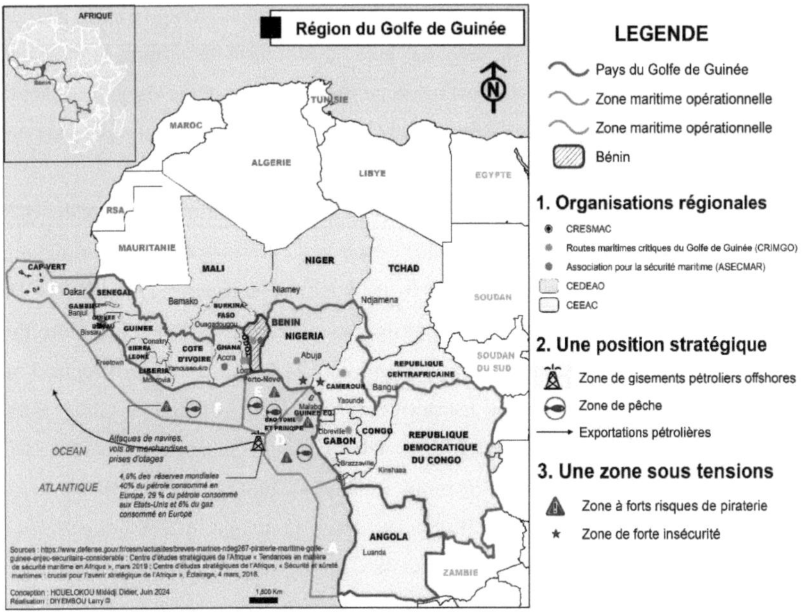

In addition, the country has "*a theoretical maritime domain appearing as a rectangle whose dimensions are equivalent to 125 km wide by 370.450 km long, i.e. 46,300 km²*".[2] This represents around 40% of the national territory. Although Benin has had a maritime code[3] since 1968, it has still not managed to control the safety of its complex maritime space. The State applies its sovereignty in a differentiated manner in maritime zones, in accordance with the 1982 Montego Bay Convention[4] : 12 nautical miles (22.22 km) of territorial sea, 188 nautical miles (348.176 km) of exclusive economic zone[5] and a continental shelf whose extent is at least equal to that of the economic zone, i.e. 200 miles.[6] The contiguous zone no longer has any legal significance following the adoption of the Montego Bay Convention.

Following the adoption of its new Maritime Code in 2011, Benin has continued to put in place its institutional and operational arrangements to combat maritime insecurity. In 2013,

---

[2] Philippe NOUDJENOUME (dir), Les frontières maritimes du Bénin, L'Harmattan, Paris, 2004, p.47.

[3] Ordinance n°38/PR/MTPTPT of 18 June 1968 establishing the Dahomey Merchant Navy Code. It remained in force until it was repealed by Act no. 2010-11 of 07 March 2011 on the Maritime Code in the Republic of Benin.

[4] Also known as the United Nations Convention on the Law of the Sea (UNCLOS), it was adopted in Montego Bay, Jamaica, on 10 December 1982. Benin ratified the Convention on 16 October 1997.

[5] EEZ: Exclusive Economic Zone

[6] Philippe Noudjenoume, op.cit, p.46.

the country adopted a national strategy for maritime protection, safety and security. United Nations resolutions 2018 (2011) and 2039 (2012) encourage countries in the Gulf of Guinea region to develop national, sub-regional and regional strategies to effectively combat maritime insecurity. These resolutions are an essential guiding document for efforts to strengthen maritime safety in the region. The national maritime strategy is part of the vision set out in the Benin Alafia foresight and planning document[7] 2025 and is just one aspect of the national safety policy and strategy . [8]

Also, as part of this drive to establish an institutional environment to ensure the safety of maritime activities in Benin, mention should be made of the creation of the national authority responsible for State action at sea[9] embodied by the maritime prefecture[10] (PREMAR). This dual-named institution is a recommendation of the national strategy for maritime protection, security and safety, and is responsible for implementing State action at sea. It replaces the maritime protection, security and safety management bodies set up by decree[11] in 2007. Benin, through its maritime border shared with Nigeria and Togo, enjoys strategic access to the Gulf of Guinea, a crucial zone for regional maritime security. To protect its territorial waters and prevent maritime piracy, Benin is committed to strengthening its security arrangements. Faced with destabilising criminal and terrorist activities in the region, including piracy, drug trafficking and human trafficking, the country has put in place a comprehensive and coordinated approach to ensure the safety of its citizens and contribute to regional stability.

Purpose and scope of the study

The study aims to analyse and evaluate the measures taken by Benin to ensure maritime security in the Gulf of Guinea, a strategic region. Based on fundamental concepts, it considers security to be an essential right for any sovereign state, especially in an area facing a variety of threats. Maritime security encompasses actions to ensure the safety of sea lanes and

[7] Drafted in 2000, Vision Bénin Alafia 2025 is a forward-looking and planning document for Benin up to 2025. It sets out the country's main development directions for 2025. It was revised in 2018. A new version, the National Development Plan 2018-2025, was adopted, taking into account the necessary adjustments. On the eve of the end of this vision, a process to define a new development plan has already begun. This will have a horizon of 2060. The aim is to project Benin 100 years after its independence.
[8] Decree no. 2008-735 of 22 December 2008 approving the national security policy and strategy.
[9] Decree no. 2014-785 of 31 December 2014 on the creation, powers and operation of the national authority responsible for State action at sea. Amended by Decree no. 2017-523 of 15 November 2017 and Decree no. 2019-450 of 09 October 2019.
[10] The Maritime Prefect was appointed by decree no. 2015-566 of 06 November 2015.
[11] Decree no. 2007-621 of 31 December 2007 on the creation, composition, powers, organisation and operation of the Maritime Protection, Safety and Security Management Bodies. This decree created two main bodies: the Conseil National de Protection, de Sécurité et de Sûreté Maritimes (CNPSSM) and the Comité Technique de Protection, de Sécurité et de Sûreté Maritimes (CTPSSM). These two bodies were never fully operational until the SNPSSM was drawn up in 2013. While the CNPSSM has never met, the CTPSSM has met only 4 times out of 16 since their respective creation.

the protection of ships, crews and goods against threats. The Gulf of Guinea is renowned for its security challenges, such as piracy and illicit trafficking, making Benin, as a coastal country in West Africa, an important player in the management of these problems. Regional and international cooperation is essential to meet these cross-border challenges, while security governance requires a holistic approach, including military, police and institutional aspects, as well as the promotion of good governance and socio-economic development. Finally, the study seeks to identify Benin's security challenges in the Gulf of Guinea and to propose strategies for strengthening its maritime security arrangements.

The scope of the study covers maritime governance in Benin, including the identification of security threats in the Gulf of Guinea, an analysis of actions and initiatives to strengthen maritime security, an examination of regional and international partnerships, and an assessment of the effectiveness and shortcomings of Benin's current security arrangements.

Aim of the study

The main objective is to examine the various aspects of maritime security in the Gulf of Guinea, focusing on Benin's role in regional stabilisation. This involves a discussion of the importance of international and regional cooperation to effectively address security challenges in this strategic area. To this end, several specific objectives have been derived from the general objective

- To analyse the effectiveness of the security arrangements and the operational capabilities of Benin's defence and security forces in terms of surveillance and protection of its territorial waters;
- Assess the maritime safety measures currently in force in Benin, including policies, regulatory frameworks and operational arrangements;
- Examine Benin's regional and international cooperation on maritime security in the Gulf of Guinea and assess its impact on the national security system;
- Propose practical recommendations aimed at strengthening Benin's capacity to meet the challenges of maritime safety in the Gulf of Guinea region.

Interest in the subject

The study on Benin's security arrangements in the Gulf of Guinea is of interest in several respects

- ❖ **Scientific interest:** Protecting Benin's maritime space is essential for preserving regional marine biodiversity, as Benin's coastal waters are home to various marine ecosystems, such as coral reefs and mangroves, which are vital for many marine species.

❖ **Strategic interest**: The Gulf of Guinea is of crucial importance in terms of maritime security due to its major economic impact and the presence of multiple threats such as piracy, drug and arms trafficking, and illegal, unreported and unregulated (IUU) fishing.

❖ **Political interest:** The protection of Benin's maritime space is of political importance because of its role in the country's territorial sovereignty. It is also of interest to the community as a whole, as it encourages regional cooperation, preservation of the marine environment and maritime safety.

Issues

Benin faces major challenges in securing its maritime space, notably due to a lack of means and resources, limited regional cooperation, a lack of coordination between stakeholders and a lack of maritime culture among those in power. To guarantee the safety of its waters, the country needs to strengthen its capacities and cooperation with regional players. Citizen participation in the debate on maritime safety is essential, as this responsibility lies not only with the State but also with individual citizens[12] . To this end, the central question of our study is formulated as follows:

How effective is Benin's maritime security strategy in countering illegal acts in the Gulf of Guinea?

This central issue raises a number of subsidiary questions:

What are the challenges and opportunities for enhanced regional cooperation in the fight against these threats?

How could Benin's security arrangements be improved to better meet future security challenges in the Gulf of Guinea?

Assumptions

Hypothesis 1: Benin's maritime security strategy faces a number of challenges in countering piracy, illicit trafficking and illegal fishing in the Gulf of Guinea. Despite the efforts made, particularly in terms of capacity building and regional cooperation, these illegal activities persist due to a lack of resources, limited coordination between national and regional players

[12] Maurice Kamto, in "*L'urgence de la pensée, réflexions sur une précondition du développement en Afrique*", éditions Mandara, Yaoundé, 1993, states on p.35 that we must "think so that we are not strangers to the city; so that nothing in society is foreign or indifferent to us; so that the State is ours, because politics is everyone's business. For, as Pericles said in the admirable funeral oration given to him by Thucydides: *"Only we regard anyone who is uninterested in the affairs of state not as a citizen at ease, but as useless"*.

and alleged acts of prevarication. To improve the effectiveness of its strategy, Benin must step up its efforts in terms of capacity building and resource mobilisation, regional cooperation and the fight against corruption.

Hypothesis 2: The challenges for enhanced regional cooperation in the fight against piracy, illicit trafficking and illegal fishing in the Gulf of Guinea include divergent priorities and approaches between countries, political tensions hampering cooperation, financial and technological shortcomings and conflicts over maritime borders. On the other hand, increased cooperation makes it possible to share information and best practices, coordinate operations, mobilise collective resources and strengthen institutional capacities. It also contributes to regional stability, fostering economic development and collective security.

Hypothesis 3: To improve maritime safety in the Gulf of Guinea, Benin needs to strengthen its security capabilities, step up regional cooperation, invest in maritime surveillance, combat corruption, strengthen the legal framework, raise awareness and mobilise citizens and adequate resources. By combining these measures, Benin could better respond to the challenges of maritime security in the region.

Methodological framework

In this work, the methodology is essentially based on a review of the existing literature on maritime safety in Benin and the Gulf of Guinea. We examined the conventions ratified by Benin, in particular the 1982 Montego Bay Convention on the Law of the Sea. We also consulted newspaper articles, analyses, dissertations and research papers to enrich our understanding of the subject

- A three-part questionnaire was also used to support some of our arguments;
- Collection of data such as statistical data on security incidents or activities in the Gulf of Guinea, interviews with experts[13] in maritime security;
- Analysis of data to identify trends, gaps and challenges in Benin's security system;
- Interpretation of the results to identify areas where Benin's security system can be improved, the factors that contribute to its effectiveness and the challenges it faces.

Coordination of work

Finally, our analysis is structured in two parts. In the first part, we diagnose Benin's maritime security organisation. After showing how the security arrangements in place are poorly adapted to the wide-ranging threats Benin faces on its maritime territory (Chapter I), we

[13] The Maritime Prefect made a valuable contribution to the production of this report. Other officers of the French Navy also contributed their expertise to help us better understand certain concerns.

examine the response mechanism for dealing with maritime insecurity (Chapter II). The second part looks at the challenges and prospects for implementing Benin's maritime strategy. The challenges are multidimensional and multifaceted (Chapter III). As for the outlook, it looks reassuring if efforts are concentrated on ensuring the coherence of Benin's maritime strategy (Chapter IV).

PART ONEDIAGNOSIS OF BENIN'S MARITIME SAFETY ORGANISATION

In a world where maritime security challenges are increasingly complex and diverse, it is essential to critically assess the maritime security organisation of a country like Benin. As a coastal state with access to the Atlantic Ocean, Benin faces a range of maritime threats, from piracy and illicit trafficking to maritime pollution.

Even if all the specialist analyses put forward the idea of a regional security architecture to form a common front against the threats facing the States of the Gulf of Guinea, it remains true that the security of a State and its border areas is primarily a national responsibility. Cooperative initiatives remain complementary practices to the national system. They are not intended to replace national security responsibilities, nor to become the systematic recourse to which States should refer.

By adopting a national strategy for maritime protection, security and safety, the aim of which is to "*reaffirm the authority of the State at sea*"[14] , Benin intends to exercise full sovereignty over its maritime domain in order to make the most of it. In this section, we will take an in-depth look at Benin's current maritime security organisation, analysing its strengths and weaknesses. We will seek to understand how Benin is tackling the challenges of maritime security. The reassertion of the State's authority at sea is based on a security system that is poorly adapted to threats from all directions (CHAPTER I). An examination of the mechanism for responding to insecurity in Benin's maritime space (CHAPTER II) will provide a clearer picture of the problems with this system, with a view to considering improvements likely to enhance maritime safety in Benin's territorial waters.

[14] National maritime protection, safety and security strategy document, p.68

CHAPTER I: A SECURITY SYSTEM THAT IS POORLY ADAPTED TO THREATS FROM ALL DIRECTIONS

Faced with the complex challenges of maritime safety, Benin's security arrangements are inadequate. While illegal activities at sea such as piracy, armed robbery, illegal transhipment, pollution and illegal fishing are on the increase, the regulatory framework and structures responsible for maritime safety in Benin seem to be struggling to respond effectively to these challenges.

In this study, we will examine in detail the gaps and weaknesses in Benin's security arrangements in the face of these all-encompassing threats, highlighting the dysfunctions in the regulatory framework, the operational shortcomings and the structural challenges that compromise maritime safety in Benin's territorial waters.

Section 1: The regulatory framework and the players involved.

The regulatory framework and the players involved in maritime safety management in Benin comprise several essential elements. On the one hand, there are laws and decrees that define the responsibilities, competencies and procedures to be followed in the maritime field. These include the Maritime Code, international conventions ratified by Benin, and specific decrees governing the various aspects of maritime safety.

1.1: The regulatory framework.

In order to regulate the national maritime space and related activities, in 2007 the Beninese Executive adopted a decree on the creation, composition, powers, organisation and operation of the Maritime Protection, Safety and Security Management Bodies. The National Council for Maritime Protection, Safety and Security (CNPSSM) and the Technical Committee for Maritime Protection, Safety and Security (CTPSSM) are the two bodies created by these regulations. The CNPSSM is responsible for[15] :

- to initiate and lead all discussions and actions likely to contribute to the proper management of maritime protection, safety and security in Benin;
- to provide all the political guidelines necessary for safety, security and the safeguarding of sovereignty and the protection of national interests at sea.

[15] Article 2 of Decree no. 2007-621 of 31 December 2007 on the creation, composition, powers, organisation and operation of the Maritime Protection, Safety and Security Management Bodies.

The CTPSSM is responsible for assisting the CNPSSM.[16] This regulatory act demonstrates the political will to take control of the sea and associated activities. However, the bodies set up by decree in 2007 have not really functioned. Composed mainly of ministers and directors of administrations involved at sea, they have been paralysed by the overloaded agendas of these senior civil servants.

Faced with the resurgence of insecurity in its maritime space from 2009 onwards, it became urgent to act in order to provide Benin with a more dynamic and operational legal architecture on the issue of maritime safety.

Drawing lessons from the shortcomings of the 2007 regulatory framework, the country updated its Maritime Code in 2011. This takes into account Benin's accession to international conventions and treaties, in particular the United Nations Convention on the Law of the Sea.

The other texts adopted and which constitute the regulatory architecture of Benin's maritime security system are :

- Decree no. 2013-551 of 30 December 2013 adopting the National Strategy for Maritime Protection, Safety and Security;
- Decree no. 2014-785 of 31 December 2014 on the creation, organisation, powers and operation of the National Authority in charge of State Action at Sea;
- Decree no. 2020-270 of 06 May 2020 requiring commercial vessels bound for Benin ports to be armed;
- interministerial order[17] n°2020-016 on the protection of ships in Benin's territorial waters.

To this regulatory framework should be added the remit of the French Navy, the operational arm responsible for implementing the security arrangements. Broadly speaking, the legal framework governing Benin's maritime environment is as follows:

[16] Ibid. Articles 8 and 9.

[17] This order implements Decree no. 2020-270 of 06 May 2020 requiring commercial vessels bound for Benin ports to have an armed guard. In accordance with article 1[er] of this decree, "all ships bound for a Benin port are required to have an armed on-board protection team (EAPE). Failing this, it is compulsory for the Beninese public forces to provide protection when entering Benin's territorial waters. The related costs are paid to the port of call.

Figure No.1 : Legal and regulatory organisation of Benin's maritime area.

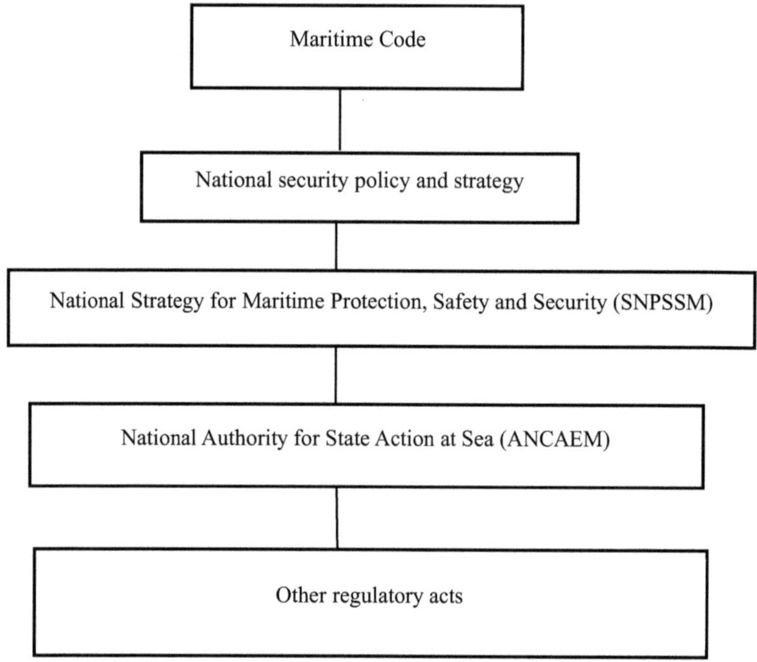

Source: Field data, 2024 Produced by HOUELOKOU, 2024.

This diagram, which is in line with the hierarchy of standards, is sometimes illegible in practice. Made up of scattered and sometimes hard-to-find texts, the players responsible for its implementation are not as obvious to those outside the maritime sector.

1.2: The players involved.

To make this legal arsenal work, various players are involved at different levels, and their effective involvement is necessary to achieve the desired effectiveness. The first of these players, the one who assumes the leadership functions and on whom rests the reaffirmation of the State's authority at sea, is the Maritime Prefect. Appointed by decree in the Council of Ministers, the maritime prefect embodies the national authority responsible for State action at sea (ANCAEM) and is vested with "*administrative policing powers at sea. He controls the conditions for the use of force at sea*".[18] It implements the top-down management approach.[19]

As shown in the figure below, the players involved in government action at sea in Benin are hierarchically structured.

Figure No.2 : Organisational chart of the institutional framework for government action at sea. [20]

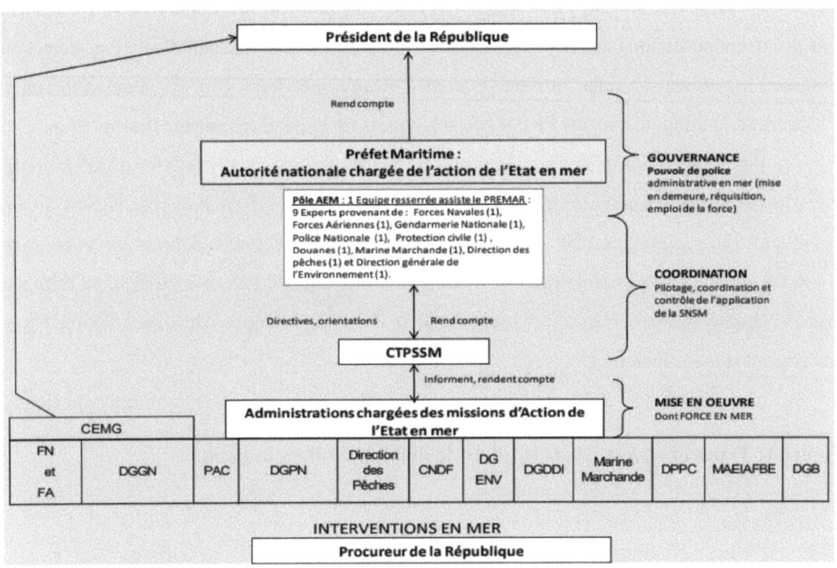

Source: SNPSSM, 2013.

As a result of the various revisions[21] to the Beninese Criminal Code in 2013, the national courts, in particular the Court for the Repression of Economic and Terrorist Offences (CRIET)[22] , now have jurisdiction to try people suspected of acts of piracy. In addition, officers

---

[18] National maritime protection, safety and security strategy document, p.72.

[19] This is a managerial approach in which decisions are taken at the highest level of the hierarchy, who then give orders to the lower echelons to carry them out. This is precisely the case with PREMAR, which is vested with national authority for state action at sea. The problem with this management model is that the administrations involved do not always work together optimally.

[20] Since the adoption of the SNPSSM document in 2013, a number of changes have taken place. These include the merger of the National Police and the National Gendarmerie into the Republican Police in 2018, and the creation of the Court for the Repression of Economic Offences and Terrorism (CRIET) in 2018.

[21] Law no. 2012-15 of 18 September 2013 on the Criminal Code in the Republic of Benin has been amended and supplemented by the following successive laws: law no. 2018-14 of 14 February 2018; law no. 2020-23 of 29 September 2020; law no. 2022-19 of 19 October 2022 and law no. 2022-37 of 20 December 2022. These various amendments to Benin's Criminal Code reflect the legislature's desire to take account of changes in the overall security situation and the need to adapt the national judicial system to deal with them.

[22] Act No. 2008-13 amending and supplementing Act No. 2001-37 of 27 August 2002 on the organisation of the judiciary in the Republic of Benin, as amended, and establishing the Court for the Repression of Economic Offences and Terrorism.

commanding ships of the French Navy, as well as second-in-command, have been empowered as judicial police officers (OPJ) to "*record acts committed at sea by means of reports that are authentic until proven otherwise. These reports are immediately forwarded to the competent prosecutors.*" [23]

In practical terms, the implementation of this legal framework begins with action by the French Navy. It is the Navy that triggers government action at sea, either by a ship on patrol that discovers suspicious activity, or by the surveillance centres (semaphore)[24] that detect an abnormal movement on radar or receive an alert from a merchant ship. As soon as the navy discovers something, it informs PREMAR, which sets up its staff to manage the situation.

For intelligence gathering, Interpol or special services can provide information on specific cases. The Yaoundé architecture can also provide information on vessels to be monitored. On the commercial side, the Merchant Navy Directorate defines the anchorage zones for ships waiting to enter the port of Cotonou. Close coordination is established with the French Navy to validate these zones, depending on the ship protection capacities that the Navy can deploy at these locations.

The graph below shows the types of stakeholders involved in maritime safety in Benin.

Figure 1: Types of stakeholders involved in maritime safety in Benin.

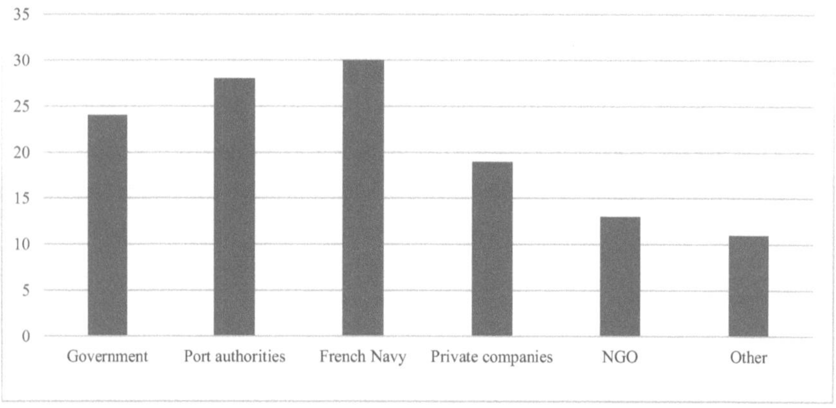

Source: Field data, 2023. Based on the questionnaire in Appendix 1.

[23] Article 29-1 of Act no. 2020-23 of 29 September 2020.

[24] The semaphore is a point of communication with ships approaching ports. It is to the navy what the control tower is to airports

As diverse as they are, these different players carry out a variety of activities at sea. As a result, they do not pursue the same objectives but are subject to the same vulnerability: the safety of the sea depicted in their activities.

In summary, the State's action at sea is based on a regulatory framework that is constantly evolving at national level, in order to respond to emerging threats in the waters under Benin's jurisdiction. This legal and regulatory framework is implemented by clearly defined players, with PREMAR occupying a central position as the leading authority, with the French Navy as the operational entity.

In response to the security challenges in the Gulf of Guinea, the United Nations Security Council adopted resolutions 2018 (2011) and 2039 (2012), inviting the countries concerned to develop national, sub-regional and regional strategies to effectively combat maritime insecurity.

Section 2: Manifestations of insecurity in Benin's maritime area.

"The Gulf of Guinea, which stretches from Senegal to Angola, has become the third high-risk maritime zone, along with South-East Asia and the Gulf of Aden.[25] In response to this security challenge, the United Nations Security Council adopted resolutions 2018 (2011) and 2039 (2012), inviting the countries concerned to develop national, sub-regional and regional strategies to effectively combat maritime insecurity.

Maritime insecurity, as defined in this work, encompasses a multitude of acts on Benin's maritime territory. These acts manifest themselves in various ways, ranging from maritime piracy to armed robbery, illegal transhipment, pollution, overexploitation of maritime resources and illegal, unregulated and unreported (IUU) fishing. During 2011, and particularly between April and July of that year, the security situation at sea deteriorated significantly.

2.1: Maritime piracy.

Sea piracy is defined as a criminal act committed at sea, involving the use of force or the threat of force, by individuals or groups for personal or economic gain. These acts include stealing, boarding, hijacking or taking control of ships, as well as capturing hostages on board. Maritime piracy may also involve other illegal activities such as drug trafficking, arms trafficking or human trafficking.

[25] Thierry Vircoulon and Violette Tournier, "*Sécurité dans le golfe de Guinée : un combat régional*", Politique étrangère no. 3/2015, p161-174, accessed at www.cairn.info on 24/03/2024.

Article 101 of the United Nations Convention on the Law of the Sea defines piracy as any of the following acts:

a) any unlawful act of violence or detention or any depredation committed by the crew or passengers of a ship or private aircraft, acting for private purposes, and directed :
   i) against another ship or aircraft, or against persons or property on board, on the high seas;
   ii) against a ship or aircraft, persons or property, in a place outside the jurisdiction of any State;
b) any act of voluntary participation in the use of a ship or aircraft, where the perpetrator is aware of facts from which it follows that the ship or aircraft is a pirate ship or aircraft;
c) any act intended to incite or facilitate the commission of the acts defined in points 1 a) or b).

The act constituting piracy can only take place on the high seas or in any other place outside the jurisdiction of any State.[26] Legally speaking, "*acts committed in the territorial waters of a State cannot be classified as piracy, insofar as they occur in an area under the sovereignty of a State, which alone is competent to repress them. These acts are referred to as brigandage and defined in the Code of Conduct for the Investigation of the Crimes of Piracy and Armed Robbery against Ships in Resolution 1.1065 (26) of the International Maritime Organisation, which defines it as "[...an unlawful act of violence or detention or any depredation or threat of depredation, other than an act of piracy, committed for private purposes against a ship, or against persons or property on board, in the internal waters, archipelagic waters or territorial sea of a State*".[27] In 2023, of the 31 incidents recorded in the Gulf of Guinea, 5 were piracy-related and 26 were acts of banditry.[28] Benin is therefore confronted with acts of banditry in its waters. What are they?

2.2: Acts of banditry in Beninese waters.

In recent years, acts of piracy in the Gulf of Guinea have fallen relatively, but acts of banditry have remained stable. This diagram summarises the situation.

[26] Article 100 of UNCLOS.
[27] MICA Center, 2023 annual report.
[28] Ibid.

Figure 2: Number of events in the Gulf of Guinea in 2023

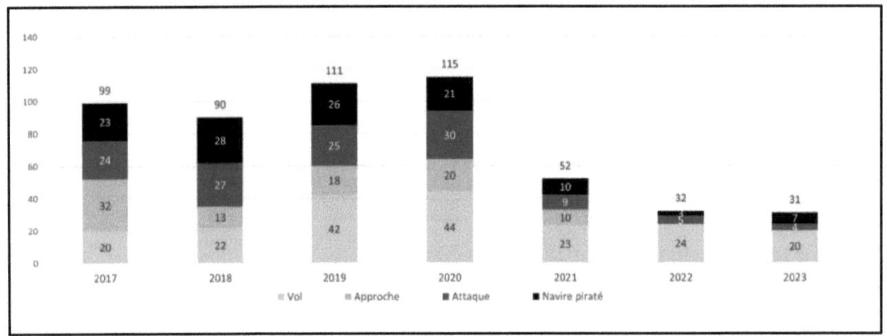

Source: MICA Center, Bilan 2023, p.27.

This graph shows that 2020 saw the highest number of incidents in the Gulf of Guinea. Theft is the highest illegal act among these incidents; attacks recorded during the same year in this area are also the highest over the period.

The situation regarding acts of banditry at national level is as follows:

Figure 3: Number of events in Benin from 2008 to 2020.

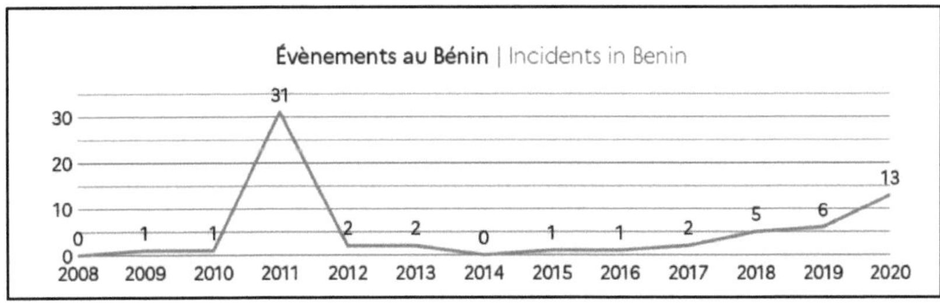

Source: MICA center, op.cit.

Most of these acts involve kidnapping crew members for ransom, hijacking vessels to serve as mother ships, cargo theft and bunkering, theft (deck fittings, crew's personal belongings), illegal fishing, illegal immigration, arms and drugs trafficking, illegal dumping and pollution.

Although the incidence of maritime crime is becoming less and less frequent thanks to measures taken by the national authorities, in particular the requirement for ships bound for

Benin ports to have an armed protection team on board[29] , waters under Beninese jurisdiction are still not spared from this form of maritime crime.

With jihadist terrorism a threat that Benin's armed forces have been facing on the country's northern front since 2021, it is easy to understand that cross-border maritime crime could be relegated to the background and receive less attention from the public authorities.

The fact that these acts of banditry are less frequent does not mean that the threat has disappeared, so isn't it time to examine Benin's response mechanism to maritime insecurity?

Plate No.1 : Routine activities of the Benin Navy.

Photo N°1 : Internal exercise of the French Navy	**Photo N°2** : French Navy nautical vector on patrol

Source: French Navy Operations Division, 2023.

As these photos show, the French Navy is committed to its missions. In fact, these various exercises are designed not only to ensure a visible presence of the force at sea, but also to maintain the know-how acquired by the personnel during their various training courses. Exercises such as these are necessary, and when they are organised regularly, they help to dissuade potential criminals operating at sea and keep personnel alert.

[29] Article 1er of interministerial order n°2020-016/MIT/MDN/MSP/MEF/DC/SGM/CJ/SA/020SGG20 on the protection of ships in Benin's territorial waters.

CHAPTER II: EXAMINATION OF THE RESPONSE MECHANISM TO MARITIME INSECURITY IN BENIN.

The review of Benin's response mechanism to maritime insecurity is of crucial importance in the current context of regional and global security. Maritime security is of growing importance in the Gulf of Guinea region, which is facing an increase in acts of piracy, armed robbery at sea and other forms of transnational crime. In this context, Benin, as a coastal State, must have an effective response mechanism to protect its national interests at sea, ensure the safety of its waterways and contribute to regional stability.

This chapter therefore aims to analyse in depth the strengths, weaknesses and challenges of Benin's current response mechanism to maritime insecurity, highlighting the various institutional, operational, technological and financial aspects that underpin it. By gaining a better understanding of the gaps and needs of the current mechanism, it will be possible to formulate strategic recommendations to strengthen Benin's capacity to meet the challenges of maritime safety in the region.

So what are the shortcomings of Benin's response to maritime insecurity (Section 1) and how does the national security approach fit in with the sub-regional framework (Section 2)?

Section 1: Shortcomings of Benin's response mechanism to maritime insecurity.

The shortcomings of Benin's response mechanism to maritime insecurity raise major concerns about the country's ability to protect its interests at sea. Faced with a growing threat in the Gulf of Guinea region, an in-depth analysis of these shortcomings is essential to guide efforts to strengthen Benin's maritime security

1.1: Factors explaining the inadequacy of the response mechanism.

To understand this concern, we submitted a questionnaire to some officers of the French Navy. While it should be noted that those questioned were able to identify precisely the types of players involved in maritime safety in Benin, they nevertheless felt that responses to maritime incidents were of average speed and that the resources allocated were inadequate.

The speed of response in the event of a maritime incident is a key factor in mitigating risk and minimising damage.

Figure 4: Speed of response in the event of a maritime incident.

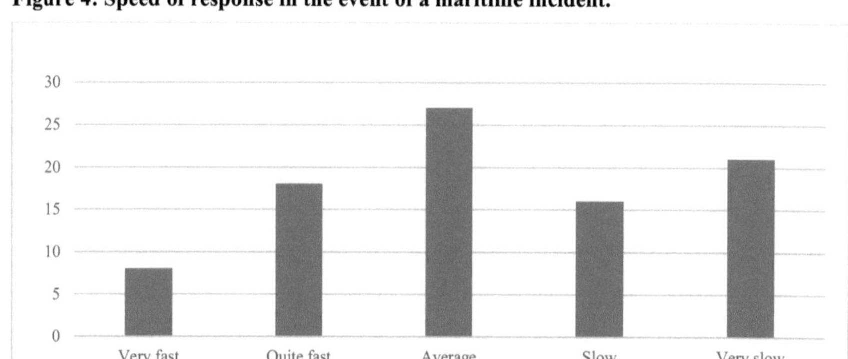

Source: Field data, 2023. Based on the questionnaire in Appendix 1.

The speed of response in the event of a maritime incident is considered to be average because of the inadequacy of the resources allocated. In other words, for those questioned, if the resources allocated were sufficient, the response would have been faster.

As for the resources allocated, their adequacy plays a crucial role in ensuring a rapid and effective response in the event of a maritime incident.

Figure 5: Sufficiency of resources allocated.

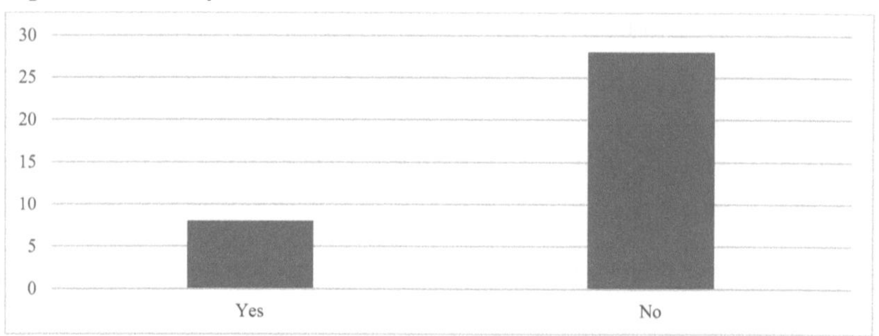

Source: Field data, 2023. Based on the questionnaire in Appendix 1.

The resources allocated for response in the event of a maritime incident are considered to be largely insufficient by those questioned.

The answers they gave to the categories of missing resources and to the question of whether or not the training offered to maritime safety stakeholders was adequate complete our understanding of the control of the response mechanism. Indeed, 100% of those questioned felt that the missing resources were primarily material. They are also financial and human, as shown in the graph below.

Figure 6: Categories of missing resources.

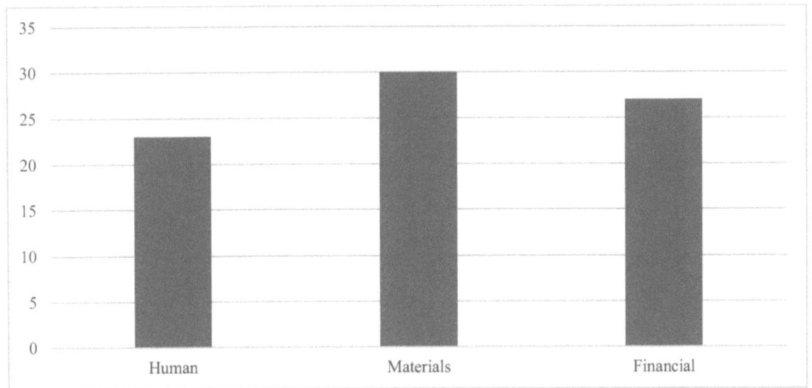

Source: Field data, 2023. Based on the questionnaire in Appendix 1.

The response to maritime incidents in Benin is said to be of average speed due to the insufficient material and financial resources allocated to the French Navy, which is responsible for implementing this response.

The personnel of this force are perceived as having received adequate training to react correctly in the event of a maritime incident. There is therefore no problem of competence in this respect. This is shown in the following graph:

Figure 7: Adequacy of player training.

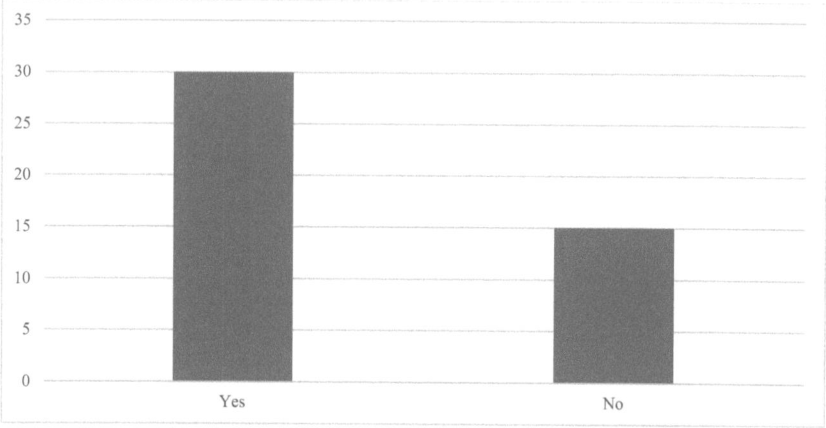

Source: Field data, 2023. Based on the questionnaire in Appendix 1.

This graph shows that the personnel of the French Navy are presumed to have received adequate training to carry out their mission properly. The Navy's ineffectiveness in dealing with certain maritime incidents cannot therefore be attributed to a training problem.

In addition to these explanatory factors, an induced analysis can be carried out to identify the shortcomings of Benin's response mechanism to maritime insecurity.

1.2: Induced analysis.

In the diagnosis drawn up by the National Strategy for Maritime Protection, Safety and Security (SNPSSM) in 2013, it was noted that national capabilities to effectively combat maritime threats were insufficient. A decade later, the situation appears to be stagnating, as revealed by the responses provided by participants to the questionnaire entitled "Identification of weaknesses in the response system". They believe that the inadequacies of the response system are attributable, on the one hand, to the inadequacy of the resources allocated and, on the other hand, to the technological challenges that would moderately affect this system.

The lack of coordination between those responsible for implementing the response system is also seen as an obstacle. When a French Navy patrol detects suspicious activity at sea, it immediately reports to its military hierarchy at . Depending on the nature of the suspicious activity, a team of representatives from the various administrations involved at sea is formed around the PREMAR. This staff then takes command of operations in the management of this particular event. The staff of the French Navy relay the orders to be implemented by the patrols at sea.

To sum up, the shortcomings of Benin's response mechanism to maritime insecurity are due both to the technological challenges faced by the equipment available and to the lack of coordination between the players responsible for implementing the response mechanism.

It is important to stress that the legislative framework is perceived as being fairly effective. Indeed, since ship captains have been empowered as judicial police officers[30] (OPJ), it has become easier to detect offences and apprehend suspects. This police power now begins at sea and continues on land.

Figure 8: Perception of the legislation.

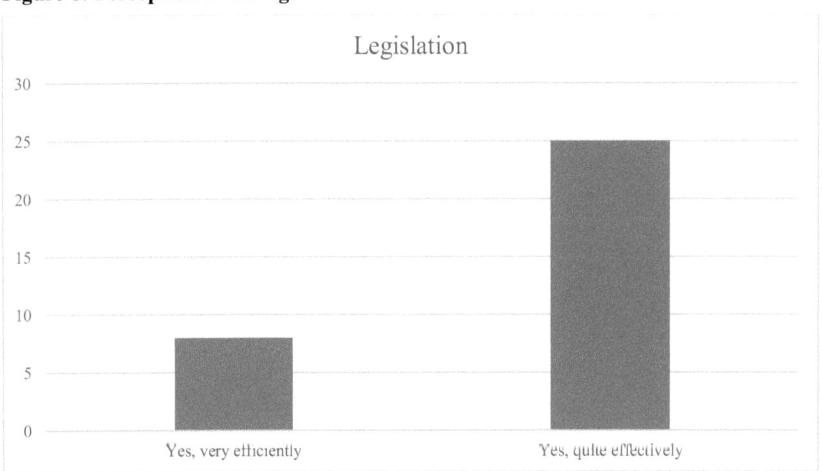

Source: Field data, 2023. Based on the questionnaire in Appendix 1.

This perception of the legislation as effective is an asset that should be capitalised on. It reassures those working at sea that successful legal action will now be taken against suspects caught at sea, unlike in the past when suspects were guaranteed to get out of police custody for lack of adequate legislation.

The national maritime safety system and the sub-regional architecture are closely linked. Benin's system is integrated into the sub-regional architecture through close cooperation and coordination between the various national entities and regional structures. This ensures a coherent and complementary approach to managing maritime safety challenges, aligning

[30] Article 29-1 of Act no. 2020-23 of 29 September 2020 amending and supplementing Act no. 2012-15 of 18 March 2013, as amended, on the Code of Criminal Procedure in the Republic of Benin.

national priorities with regional objectives and initiatives. Ultimately, this strengthens Benin's ability to contribute effectively to maritime safety in the Gulf of Guinea region.

Section 2: Benin's approach to security within the sub-regional framework.

By asserting that the Gulf of Guinea region is theoretically an important area for integration, Abdelhak Bassou[31] highlights the idea that the cooperative approach is a necessity for the States bordering the Gulf of Guinea if they wish to properly initiate the maritimisation of their respective economies. With this in mind, these States have decided to combine their security efforts within the framework of a common body: the inter-regional coordination centre[32] (CIC), which brings together the three regional communities, namely ECOWAS, CEAAC and CGG.

So how does this regional security architecture work, and what role does Benin play in this cooperation?

2.1: How the regional security architecture works

The CIC is designed as a strategic driving and coordinating body, organised on several levels. At local level, it comprises national Maritime Operations Centres (MOCs), subdivided into different zones managed by Multinational Coordination Centres (MCCs). At regional level, there are the Regional Maritime Safety Centres (CRESMAC for Central Africa and CRESMAO for West Africa). Finally, at inter-regional level, there is the CIC[33] . This centre, which is not an additional level of command in the regional security architecture, is positioned more as a facilitator of cooperation initiatives. When an incident occurs at sea, involving for example two States bordering the same area, the Multinational Coordination Centre takes charge of the intervention procedures of the operational structures of the States concerned by sharing up-to-date information on the incident in progress.

In practical terms, the effectiveness of the Information and Coordination Centre (ICC) depends on national operational capabilities. Indeed, the credibility of the ICC's actions is closely linked to the effectiveness of the operational capabilities of the various entities responsible for safety at sea in the coastal states.

In other words, apart from its leadership activities, "*strengthening, sharing experience, collecting and disseminating information, coordinating the activities of CRESMAC and CRESMAO, promoting the harmonisation of legislation, developing the harmonisation of*

[31] ABDELHAK Bassou, "La mer du golfe de Guinée : richesses, conflit et insécurité", revue maroco-espagnole de droit internationales et relations internationales, no. 2, 2014.

[32] Born of the Yaoundé process, the CIC is the culmination of the political will of the States of the Gulf of Guinea to pool their security efforts to deal with maritime threats.

[33] Mathieu ILINCA, "le CIC, clé de voûte de l'architecture de coopération interrégionale dans le golfe de Guinée", Revue Défense nationale 2016/7 (N°792), p.93-98, consulted on www.cairn.info on 17 March 2024.

standard operating procedures"[34] , the ICC will only be seen as a credible body when the national navies of the riparian states have real capabilities that enable them to cooperate to succeed in their security missions.

The figure below highlights the importance of regional cooperation in tackling maritime insecurity in the Gulf of Guinea. It also highlights the relevance of analysing the specific role of each state in this cooperative approach. This is precisely what we will focus on in the following paragraph .

The regional security architecture is organised as follows:

Figure No.3 : Yaoundé architecture organisation chart

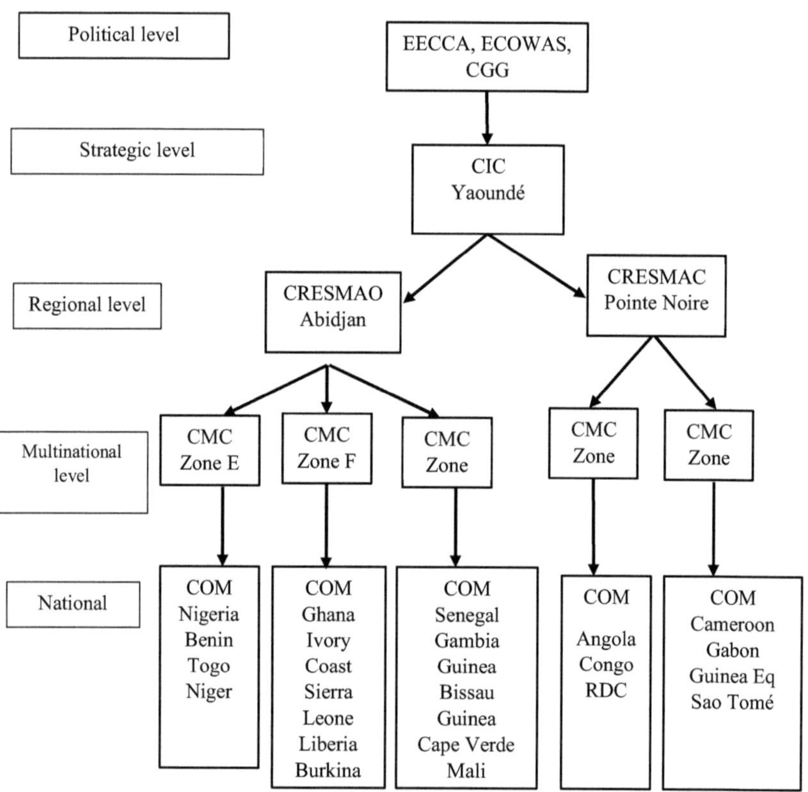

Source: MARITIMAFRICA, 2020.

[34] Ibid.

Exercise OBANGAME 2024 is a crucial multinational initiative to strengthen maritime security in the Gulf of Guinea, a strategic region facing major challenges such as piracy, drug and arms trafficking, and illegal fishing. In this context, the human and material resources deployed by the French Navy bear witness to Benin's proactive role in the regional approach to maritime security.

Plate N°2 : Human and material resources of the French Navy involved in exercise OBANGAME 2024

Photo N°3 : Medical assistance to a disaster victim	**Photo N°4 ; Operations centre**
Photo N°5 : Pirate ship boarding	**Photo N°6 : Approaching a target vessel**

Source: Field data, 2024.

The participation of the Beninese Navy in the OBANGAME 2024 exercise, through its human and material resources, demonstrates its commitment to strengthening maritime security in the Gulf of Guinea. By working alongside the navies of other countries in the region, particularly those in Zone E, Benin is demonstrating its interest in international cooperation and the promotion of regional stability, while at the same time demonstrating its operational capabilities and expertise in the management of contemporary maritime threats.

2.2: Benin's place in the regional maritime safety architecture.

Benin is committed to complying with international and regional instruments to combat insecurity off its waters, and stands out for its active participation in collective security approaches in the maritime field. Indeed, "*since the call for international solidarity launched by President Boni YAYI in the summer of 2011, which led to Security Council Resolution No. 2018 of 31 October 2011, confirmed in its recommendations by Resolution No. 2039 of 29 February 2012, which emphasise sub-regional and regional cooperation through the establishment of a strategic framework*"[35] , Benin has made a positive contribution by putting in place all the legal instruments needed to combat maritime insecurity at national level, and by playing an active role in setting up the sub-regional structures resulting from the implementation of the Yaoundé architecture. This is how the headquarters agreement between ECOWAS and the Republic of Benin relating to the Multinational Maritime Coordination Centre (CMMC)[36] in maritime zone E[37] was signed in Abuja in 2016. This centre, whose mission is to "*promote cooperation between the Parties to enable them to combat and eradicate piracy, armed robbery at sea and other illegal maritime activities more effectively*"[38] , saw the signing in August 2018 of the memorandum of understanding on the implementation of joint maritime operations and patrols between the Member States of ECOWAS maritime zone E.

As the CMMC is the lowest level of the Yaoundé architecture, it is precisely here that the role played by the coastal State in this regional initiative can be measured. Article 3 of the memorandum specifies that the maritime area covered by this memorandum "*comprises international waters and waters under the respective jurisdiction of the States of Benin, Nigeria and Togo*", and establishes a "*naval air group maritime zone E*"[39] responsible for carrying out joint operations and patrols initiated by the CMMC. The latest joint exercise to date under this

---

[35] SNPSSM, p.46.

[36] The acronyms CMC and CMMC are equivalent and can be found in specialist literature. CMMC for multinational maritime coordination centre is generally the local translation of the CMC provided for in the Yaoundé architecture.

[37] This maritime zone was established by a multilateral agreement on 15 July 2013 in Abuja.

[38] W. GBAGUIDI, "*Insécurité maritime au Bénin : réponses institutionnelle et opérationnelle pour la sécurisation du trafic maritime*", doctoral thesis from the University of Abomey-Calavi, defended publicly on 13 April 2023.

[39] Memorandum of Understanding (MoU) on the implementation of joint maritime operations and patrols between the Member States of ECOWAS maritime zone E, article 4, paragraph 6: "The air and naval units engaged by the Parties shall form a maritime zone E carrier battle group, the tactical command of which shall be provided by the coastal State".

sub-regional initiative is Operation Safe Domain II[40] . As a fully operational tool[41] , CMMC Zone E functions as follows.

Figure No.4 : Command and control levels for operations and patrols in Zone E

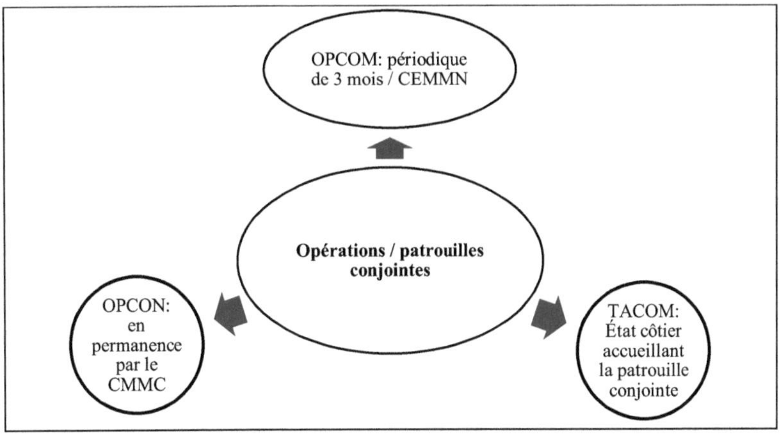

OPCOM: périodique de 3 mois / CEMMN

Opérations / patrouilles conjointes

OPCON: en permanence par le CMMC

TACOM: État côtier accueillant la patrouille conjointe

Source: MoU of 30 August 2018 Directed by HOUELOKOU, 2024.

Functioning as a complement to national security arrangements, the CMMC zone E's "*aim is to ensure permanent, joint and coordinated control of maritime zone E with a view to guaranteeing maritime safety and security*".[42] . Although this ambition is expressed periodically through joint exercises and sometimes through coordinated interventions in the event of a crisis at sea, it remains dependent on national operational capabilities. If these are weak in one State, the whole sub-region will suffer the consequences. In Zone E, only Nigeria has credible operational maritime capabilities. In this context, the proactive role played by Benin in these regional initiatives lacks visibility and credibility due to the weakness of its operational capabilities

[40] Officially launched on Monday 11 September 2023, the operation to patrol and secure the Zone E maritime area involves Togo, Nigeria and Benin under the leadership of the Zone E CMMC. It ended on Friday 15 September 2023.

[41] The exercise of military command in a multilateral framework involves different levels of command and control. This division of responsibilities is part of a system of accountability. The terms OPCOM, OPCON and TACOM refer respectively to operational command, operational control and tactical command.

[42] MoU, op.cit.

Partial conclusion

At the end of this section, the assessment of Benin's maritime security organisation highlights both progress and persistent challenges. While the country has adopted regulatory frameworks and policies aimed at strengthening its maritime security, there are still shortcomings in terms of effective implementation and coordination between the various players involved. The recent decline in acts of piracy in the Gulf of Guinea is encouraging, but the persistence of acts of banditry highlights the need for continued vigilance and appropriate measures to protect Benin's territorial waters.

To effectively strengthen the country's maritime safety, it is essential to enhance operational capabilities, improve coordination between the various bodies and promote closer regional cooperation.

By adopting an integrated approach and making a firm commitment to overcoming the challenges, Benin can better protect its interests at sea and contribute to the stability of the Gulf of Guinea region.

PART TWO: THE CHALLENGES AND PROSPECTS INVOLVED IN IMPLEMENTING BENIN'S MARITIME STRATEGY

The development and implementation of an effective maritime strategy is essential for Benin, as a country bordering the Gulf of Guinea. This strategy is designed to meet a number of multidimensional and multifaceted challenges relating to security, the economy and the environment in the maritime context.

In this second part, we explore the challenges and opportunities associated with the implementation of Benin's maritime strategy, as well as the prospects for the country's sustainable development and prosperity in the maritime sector. Benin's maritime strategy is closely linked to major geopolitical issues and crucial prospects for the country's maritime development. As a country bordering the Gulf of Guinea, Benin plays an important role in regional and international dynamics in terms of security, maritime trade and protection of the marine environment.

At a geopolitical level, implementation of Benin's maritime strategy requires effective management of relations with neighbouring countries and international players operating in the region. Maritime security issues, such as piracy, illicit trafficking and transnational crime, require enhanced regional and international cooperation to guarantee the stability and safety of Benin's waters.

CHAPTER III: MULTIDIMENSIONAL AND MULTIFACETED ISSUES.

Maritime issues occupy a central place in today's geopolitical landscape, reflecting the security, economic and environmental challenges facing nations around the world. These issues, characterised by their multi-dimensionality and variety, raise crucial questions about the security, resource management and economic development of coastal states. This chapter explores the complexity of these maritime issues, highlighting their varied nature and the challenges they pose for the international community. Through this analysis, we seek to understand the implications of these issues for regional stability and the strategies needed to address them in an effective and coordinated manner.

In an increasingly interconnected world, coastal nations face complex and evolving challenges in maritime safety and marine resource management. To meet these challenges, the development and implementation of effective policy frameworks is essential. However, adapting these strategic frameworks to the changing realities of the maritime context remains a constant challenge.

Section 1: A strategic framework that needs to be adapted.

This section looks at the need to adapt Benin's maritime strategic framework by highlighting the challenges encountered, the gaps observed and the opportunities. By analysing past experiences and best practices, we will explore possible ways of enhancing the effectiveness of maritime strategies in a constantly changing environment

So what are the shortcomings of the national maritime protection, safety and security strategy and what improvements can be made to make it more operational?

1.1: Shortcomings of the national maritime protection, safety and security strategy.

Almost 10 years after the adoption of the strategy, not all the administrations concerned have yet fully complied with the coordination requirements, and inter-administration is not fully effective. This simply reveals the lack of a 'maritime culture' among certain players in the maritime chain. The establishment of ANCAEM has a number of structural weaknesses. This authority, which was intended to be the coordinating body for all the national administrations involved at sea, was also supposed to be the body responsible for implementing the SNPSSM. However, the SNPSSM makes no provision for a separate structure responsible for monitoring and evaluation. This shortcoming in the monitoring and evaluation system shows that certain specific objectives have not been achieved.

For example, with regard to the specific objective of "*developing the capacity to control maritime space*"[43] which is to be achieved through the strategic axis of "*promoting better governance of the State's action at sea*"[44] , the activity of setting up the "*centre of experts to assist the Maritime Prefecture*"[45] has never been carried out. The Préfecture Maritime was therefore set up and is operating without the centre of experts.

This initial structural handicap has increased with the rank conferred on the Maritime Prefect. While the SNPSSM provided that the ANCAEM "*shall take the name and functions of Maritime Prefect (PREMAR) and shall have the rank of Minister*"[46] , the 2017 amending decree disregards this strategic obligation and provides that *"the Maritime Prefect shall be appointed by the President of the Republic from among the senior naval officers of the operational branch, of the rank of Captain (Navy) at least."*[47] The PREMAR has therefore been established without a reinforced authority, despite its role as coordinator to promote better governance of the State's action at sea.

Among the challenges facing the State's action at sea in Benin are the low level of support from the administrations concerned, the lack of coordination between stakeholders and the absence of standardised operational procedures. In an area as critical as maritime safety, it is imperative that all stakeholders work closely together and follow well-established protocols to ensure the safety of people and property. With regard to the strategic priority of "*modernising the State's operational means of intervention at sea*"[48] , the shortcomings and weaknesses identified include the following:

- the inadequacy and/or obsolescence of the resources acquired in relation to the objectives pursued;
- insufficient operational equipment ;
- the absence of an equipment maintenance programme;
- the decommissioning of one patrol boat, 4 speedboats and 5 inflatable boats.

Indeed, even in its international cooperation dimension, the SNPSSM has had its shortcomings. For example, the joint operation between Benin and Nigeria, called 'Prosperity' and launched in September 2011 for a period of six months, failed due to logistical problems.

---

[43] SNPSSM document, op.cit.
[44] Draft assessment document for the SNPSSM.
[45] Ibid.
[46] SNPSSM document, op.cit. This idea of the SNPSSM to make the PREMAR a minister, was enshrined in Decree No. 2014-785 of 31 December 2014 on the creation, organisation, powers and operation of the ANCAEM.
[47] New Article 8 of Decree no. 2017-523 of 15 November 2017 amending Decree no. 2014-785 of 31 December 2014.
[48] Draft assessment document for the SNPSSM.
[48] Ibid.

In this operation, Benin had operational command (OPCOM) while tactical command (TACOM) fell to Nigeria. However, Nigeria provided 95% of the logistical assistance ([49]), while Benin contributed only 5%. Nigeria committed two helicopters, two ships and two interceptor boats, while Benin provided only two US-supplied Defenders boats.

In addition, due to the absence of a national framework for consultation, the actions of Technical and Financial Partners (TFPs) in the context of the State's action at sea are often disorganised and not always aligned with national interests.

The combination of these shortcomings has made implementation of the SNPSSM ineffective. With the emergence of a new concept of blue economy in Benin, it is timely to propose some improvements that could be beneficial in the revision of the SNPSSM.

1.2: Improvements to be made to the SNPSSM.

Having a coastline and being a country bordering the Gulf of Guinea automatically gives a country a maritime vocation. Consequently, having a maritime strategy means being aware of this geographical vocation, which influences both the country's internal and external policies. A maritime strategy "*is first and foremost the recognition that there are issues at stake. It is then an ambition and an expression of power*" [50]

The Stratégie Nationale de Protection et de Sécurité de l'Espace Maritime (SNPSSM) (National Strategy for the Protection and Security of Maritime Space) meets these criteria in part: it recognises the opportunities linked to the sea that can be exploited to Benin's advantage, and it affirms a clear ambition: to reaffirm the authority of the Beninese state at sea.

However, it was not conceived as an expression of maritime power. Moreover, its heavy dependence on sub-regional and/or international means and mechanisms for its implementation clearly underlines this lack of expression of power.

So as not to repeat the same mistakes made in implementing the SNPSSM, the bodies responsible for defining or updating the new blue economy strategy could explore the following avenues:

- Aligning the national strategy with international and regional directives could effectively compromise the State's ability to assert itself fully and risk remaining under foreign influence. It is crucial to recognise that the presence of a seafront imposes a historical responsibility in maritime matters on the country. This

[49] Raphaël TIWANG WATIO and Messan LAWSON, "*Maritime piracy in the Gulf of Guinea*", Centre de Droit Maritime et Océanique, University of Nantes, vol. 20, 2014/2

[50] Hervé Moulinier, op.cit.

awareness of the importance of the maritime domain, intrinsic to Benin's geography, must be a central element in the development of any public policy, particularly a maritime strategy. With this in mind, it is imperative to organise a public debate, involving all the players concerned, including research centres and specialist universities. The aim of this debate is to inform national public opinion on the direction the country should take in maritime matters.

- The creation of a marine research centre, operating in parallel with other structures such as ANCAEM, would be essential. This centre would be tasked with carrying out in-depth studies on all aspects of the sea, enabling us to anticipate changes in the marine environment and make the most of this geographical advantage. With adequate resources and qualified staff, this research centre would be the intellectual heart of the system put in place by the national strategy.
- The new strategy will have to take into account the shortcomings observed during the implementation of the SNPSSM. In particular, it should break new ground by ensuring that the institutional chain of government action at sea is clearer. Current legislation and regulations in this area are scattered and difficult to access. By defining a single maritime regulatory architecture and making it accessible online, this new strategy would facilitate access to maritime regulations for a wider audience. In so doing, it would help to embed maritime culture in the population.

Elevating the position of Maritime Prefect to the rank of Minister, with direct subordination to the Head of State, would be an important step in strengthening and supporting global and inter-ministerial action against maritime crime. Taking into account these suggestions for improvement, it should be possible to achieve the desired effectiveness of the National Strategy for the Protection and Security of Maritime Space. However, to ensure the full success of this strategy, it is also essential to strengthen the operational framework that implements it.

Section 2: Strengthening the operational framework.

Despite the challenges posed by its limited operational resources, the French Navy remains the key player in Benin for missions at sea. Its ongoing commitment to surveillance and intervention along the coasts and surrounding maritime areas highlights its vital role in the country's maritime security.

How is the Benin Navy organised today to deal with maritime crime and, above all, to fulfil its missions? How can the Navy be strengthened to improve its performance?

2.1: The organisation and operation of the French Navy.

Targeted improvements in the areas of surveillance, fleet, maintenance and personnel training, as well as the addition of naval air assets, would significantly enhance the French Navy's operational capability. This would enable it to respond more effectively to current and future maritime challenges, while ensuring an effective presence at sea and optimising its ability to intervene when needed.

The French Navy's operational capacity is currently considered a weakness. The means of surveillance, processing and sharing maritime information, as well as the means of intervention at sea, are limited. The fleet consists of three semaphores[51] , one deep-sea patrol vessel, five coastal patrol vessels and high-speed launches with very limited range.

However, maintenance and upkeep problems compromise their presence at sea and their operational effectiveness due to a lack of budgetary and technical resources and qualified personnel. In addition, the lack of naval air assets limits the French Navy's ability to exercise effective control at sea.

The table below sets out the current strengths and weaknesses of the French Navy.

[51] The third semaphore (at the Sèmè naval base) came into service in 2024.

Table No.1 : SWOT analysis[52] of the Beninese Navy's activities in waters under its jurisdiction

	Forces	Weaknesses	Opportunities	Threats
French Navy (MN)	- MN has a legal existence; - Has a defined, regulatory organisation chart; - Political support	- Weak surveillance and control capabilities in the maritime area ; - Weak capacity for intervention and day-to-day operations ; - Insufficient naval and river bases ; - Lack of maintenance capacity for existing resources	- Favourable political context: ANCAEM, through PREMAR, is an opportunity for MN to step up its game; - Creation of the National Guard: an opportunity to step up the acquisition of equipment and showcase its expertise; - Safe domain" joint exercises are becoming a regular occurrence	- Piracy and acts of banditry, illegal immigration, IUU fishing; - Acts of terrorism from the sea.

Source: Produced by HOUELOKOU, 2024.

[52] SWOT = Strengths, Weaknesses, Opportunities and Threats, :s analysis tool used to identify the main strengths, weaknesses, opportunities and threats of a given situation or structure.

The organisational structure of the French Navy is crucial to its operational effectiveness and its ability to fulfil its missions. Composed of different entities working in close collaboration, it ensures the coordination and efficient execution of maritime operations. This in-depth analysis will highlight the components of this structure and examine their role in the overall functioning of the institution. Understanding this organisation will provide a better understanding of how the French Navy mobilises its resources to ensure the security and protection of the country's maritime interests. Broadly speaking, the organisation chart of the French Navy is as follows:

Figure No.5 : Organisation chart of the French Navy.

EMMN

Divisions

Cells

CMA

Ecoles et centres de formation

Formations opérationnelles

Unités de soutien

Naval Training Centre

Naval bases : Cotonou-Sèmè-Grand popo

Ladji Lagoon Brigade

Karimama River Brigade

French Navy workshops

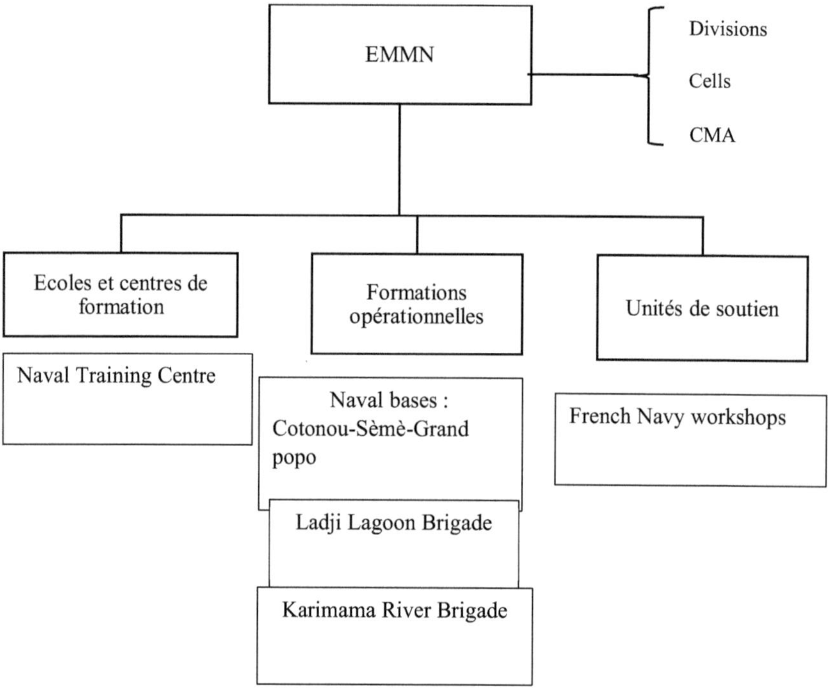

Source: Field data. Produced by HOUELOKOU, 2024.

This flexible organisation allows the French Navy to be reactive in its operations. The operational formations, commanded by naval officers with the title of Base Commander, report directly to the Chief of Staff of the French Navy. The divisions of the General Staff implement the directives of the Chief of the Naval Staff.

To carry out its assigned missions, the French Navy is deployed throughout the country. The figure below shows the French Navy's territorial network.

Figure No.6 : The French Navy's territorial network.

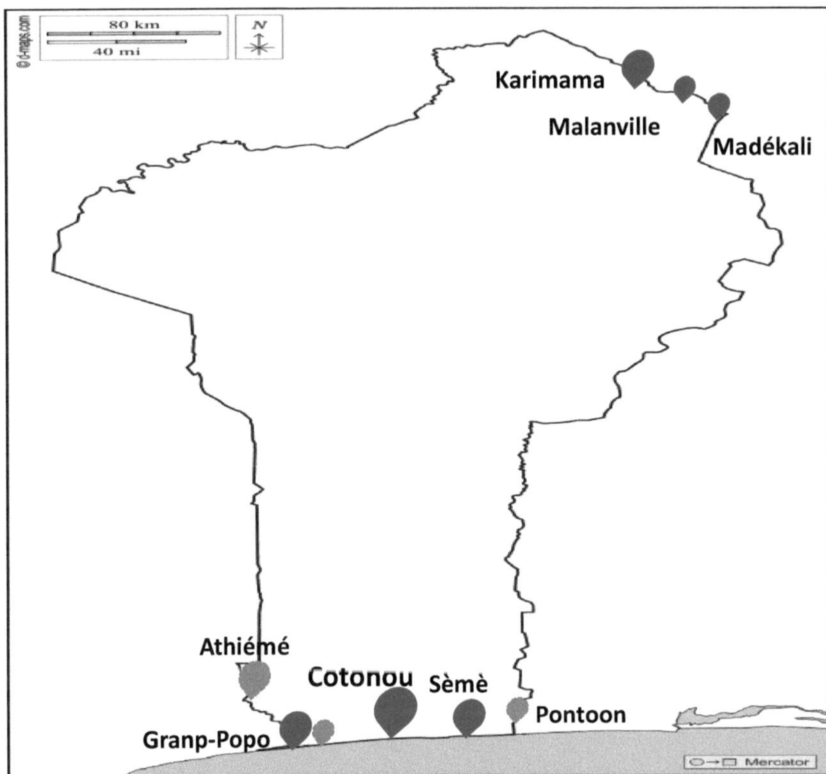

Source: Presentation by the CDS entitled "French Navy in 2024" at the 1st quarter 2024 geopolitical seminar organised by the EMG.

The naval bases are located along the coast from west to east (Gand-Popo, Cotonou and Sèmè). The northern units are located along the Niger River.

In its day-to-day operations, the French Navy carries out the regulatory missions assigned to it, particularly in terms of operational training. The Cotonou, Grand-Popo and Sèmè naval bases, with their respective semaphores, play a crucial role in this respect. As part of its participation in government action at sea, the Navy is called upon when action is triggered from the Préfecture Maritime.

Conversely, it can also take the initiative for government action at sea if suspicious activity requiring a response is detected. When this happens, a crisis unit is set up at the Préfecture Maritime to coordinate the actions of Navy units. This was the case in the April 2020 incident, where a vessel reported an attack on another vessel. In response, a French Navy patrol vessel was dispatched to rescue the vessel in distress. However, given the nature of the attack, the military hierarchy deemed the intervention of special forces necessary, which Benin did not have at that time. Consequently, the diplomatic chain was activated so that the Nigerian special forces could intervene.

Unfortunately, by the time they arrived, the criminals had already fled with the hostages.
This incident reveals the current operational capacity of the French Navy, highlighting equipment shortfalls. Nevertheless, the reforms undertaken in defence and security since 2016 have generated progress in the acquisition of major equipment. It will take a few more years to fully assess the impact of these efforts.

The French Navy relies on human resources spread across the various personnel corps in the Beninese Armed Forces, as shown in the graph below.

Figure 9: Percentage breakdown of the various corps of personnel in the French Navy.

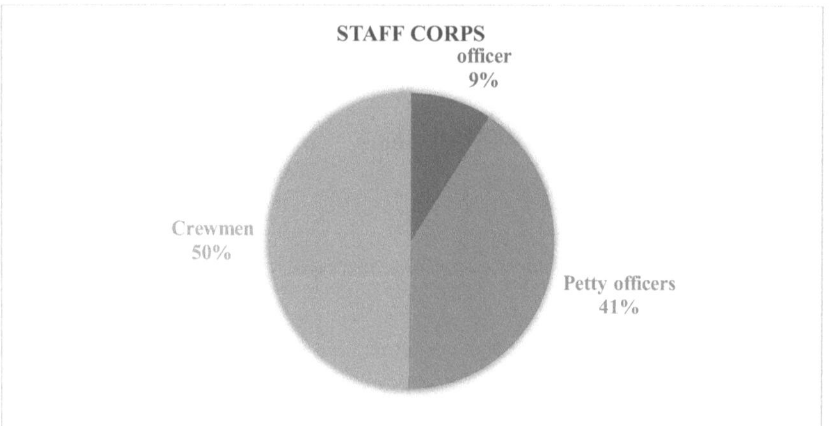

Source: Presentation by the CDS entitled "French Navy in 2024" at the 1st quarter 2024 geopolitical seminar organised by the EMG.

It is clear that the resources available to the French Navy are insufficient to carry out its missions properly and to participate in the implementation of the State's action at sea. So what improvements can be made to this force to give credibility to the nation's ambitions at sea?

2.2 : Desired reinforcements for the French Navy.

To protect the maritime zones under national sovereignty, the strategy put in place, if it is to be sustainable, must be based on a renewed availability of the French Navy's resources in order to intensify high-visibility exercises at sea, and to continue and intensify the Embedded Escort Policy (EPE) initiated by regulatory provisions. [53]

The results of the responses to the questions on proposals for improving the response to maritime threats reveal a significant perception of the direction to be taken by the security apparatus. Out of a sample of 30 respondents, 25 considered the mechanisms for international cooperation in the management of maritime security to be very effective, while 20 thought they were not very effective.

The evaluation of the effectiveness of cooperation mechanisms in the management of maritime safety reveals diverse perceptions among those questioned. This reflects varying opinions on the effectiveness of these initiatives in protecting sea lanes.

Photo N°7 : Beninese fusiliers commandos exercise with French elements

Source: French Navy Operations Division, 2021.

[53] Decree no. 2020-270 and Interministerial Order no. 016, op. cit.

Maritime safety management relies on close cooperation between various stakeholders. Bilateral agreements between neighbouring countries are essential for managing cross-border threats. For example, agreements between neighbouring countries enable joint patrol and rescue operations at sea, as well as intelligence sharing. This photo illustrates a cooperation framework between the French Navy and French elements.

Moreover, this cooperation is appreciated to varying degrees by those who responded to the questionnaire used in this study, as the following graph shows:

Figure 10: Mechanisms for cooperation in maritime safety management.

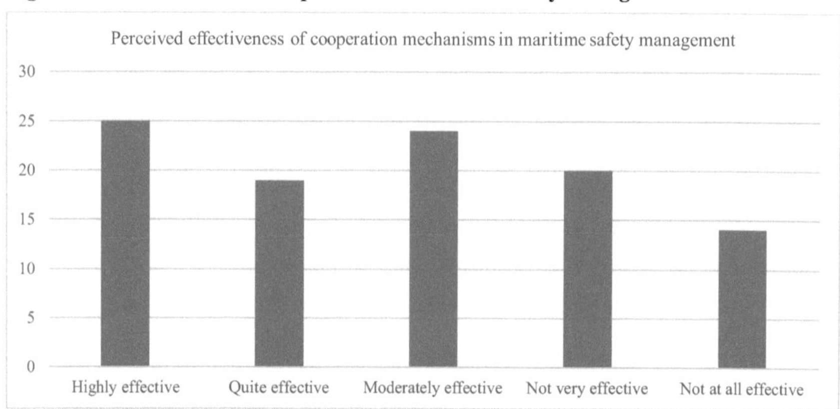

Source: Field data, 2023. Based on the questionnaire in Appendix 1.

This perception thus guides the proposals made by the respondents to improve the maritime safety system. Cooperation in the management of maritime safety is complex and multifaceted, involving a network of players and mechanisms at different levels. This cooperation is essential to guarantee the safety of maritime routes, prevent acts of piracy and illegal trafficking, and ensure an effective response to incidents at sea. The synergy between public and private players, modern technologies and international legal frameworks forms the basis of this cooperation.

In a context where maritime challenges often transcend national borders, strengthening regional maritime safety initiatives is emerging as a key priority.

Figure 11: Strengthening regional maritime safety initiatives.

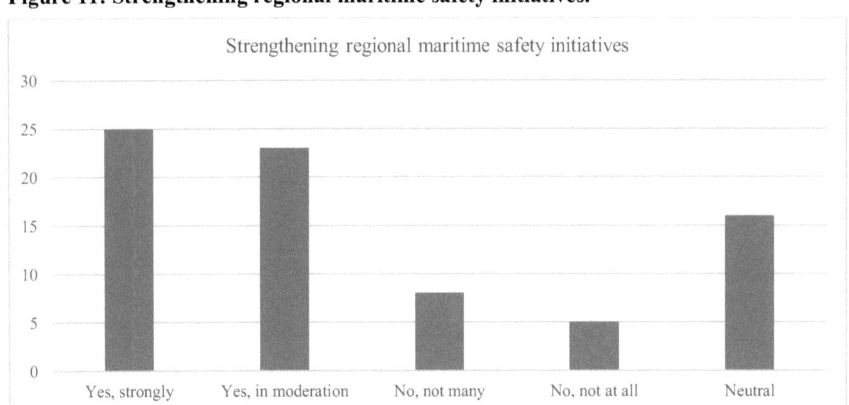

Source: Field data, 2023. Based on the questionnaire in Appendix 1.

Regional maritime security strengthens ties between neighbouring countries, encouraging cooperation and political stability. By working together on common issues, states can build mutual trust and prevent potential conflicts. Organisations such as the African Union, the GGC and ECOWAS have shown that regional collaboration can lead to lasting solutions to maritime security issues.

At the heart of concerns about maritime safety, the priority improvements to be implemented are being carefully examined to strengthen the protection of international waters.

As the following graph shows, certain improvements are desired to strengthen Benin's maritime safety system. These range from the strengthening of legislation to greater coordination between the players responsible for State action at sea, via an increase in financial resources and technological improvements.

Figure 12: Prioritised improvements to enhance maritime safety.

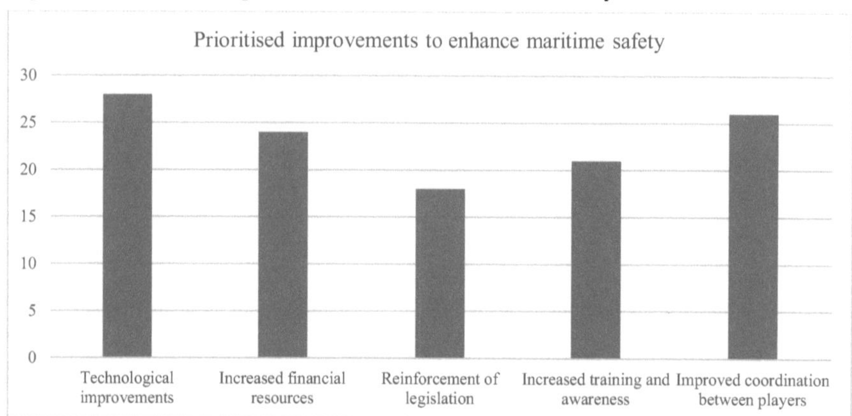

Source: Field data, 2023. Based on the questionnaire in Appendix 1.

An analysis of the graphs reveals an interesting trend: although the respondents consider the mechanisms for cooperation in maritime safety management to be very effective and would like to see regional initiatives strengthened, they also stress the importance of improving technology and increasing financial resources to further strengthen the maritime safety system. Thus, for these respondents, favouring regional or international safety mechanisms seems to be a pragmatic solution. In other words, while waiting for the desired reinforcements to become effective, it is crucial to continue to rely on regional and international cooperation initiatives, even if their effectiveness is judged to be mixed.

The major drawback of regional and international cooperation initiatives in the maritime field is the possible dependence they could induce among national players. This concern is exacerbated by the fact that the French Navy is already dependent for the upkeep and maintenance of its operational resources.

Other reinforcements for the French Navy include :

- Airborne vectors (UAVs, aircraft) for maritime patrols to increase the force's response effectiveness. However, given the small size of Benin's armed forces, it would be more appropriate to apply the principle of pooling. In this case, the Air Force would be equipped with these vectors for use by both armies.
  - Train more specialists in the various technical fields so that the maintenance and upkeep of existing and future facilities can be carried out on site.

To develop an effective maritime strategy, it is important to consider several key aspects. These include maritime safety to combat piracy, drug trafficking and smuggling, as well as the protection of waterways from pollution and other environmental hazards. In addition, a coherent strategy must address the economic challenges of maritime trade by ensuring open, efficient and safe trade routes. International cooperation is often required to develop and implement this strategy, as many maritime challenges transcend national boundaries. By working together, nations can enhance the security and stability of the world's maritime spaces, promoting effective management of marine resources and economic prosperity.

CHAPTER IV: SEEKING COHERENCE IN BENIN'S MARITIME STRATEGY

Benin has promising maritime potential for its economic development. However, to exploit this opportunity effectively requires a coherent and integrated maritime strategy. This chapter examines the various aspects of this strategy, emphasising the importance of coherence. By identifying challenges and opportunities, it proposes recommendations for strengthening Benin's maritime position and promoting sustainable development. It identifies the country's current challenges in the maritime field and explores opportunities to strengthen its position as an emerging maritime nation. Drawing on international best practice, concrete recommendations are made to improve the coherence and effectiveness of Benin's maritime strategy, with the aim of fostering sustainable and inclusive development. By adopting an integrated strategic approach, Benin can fully exploit its maritime potential and play a significant role in economic prosperity and regional security.

The existing national system for State action at sea does not comply with the principles of a maritime strategy as set out above. Hence the need to seek an effective maritime approach (Section 1). To be realistic, this need must be accompanied by the maintenance of regional and international cooperation (Section 2) in terms of maritime safety and security.

Section 1: The need for an effective maritime approach.

The need for an effective maritime approach is undeniable, especially for coastal nations such as Benin. The oceans offer countless economic opportunities, particularly in the areas of trade, fishing, tourism and energy. A well thought-out and implemented maritime strategy can not only enable these resources to be exploited sustainably, but also strengthen national security by protecting maritime borders against threats such as piracy and illicit trafficking.

1.1: A change in the way things are done.

The notion of a "break in the way things are done" implies a significant change in the way things are traditionally done. In the maritime context, this could mean abandoning obsolete or inefficient methods in favour of innovative and adaptive approaches. This break with the past may be driven by a number of factors. Firstly, developments in technology and knowledge can render old methods obsolete. For example, the use of cutting-edge technologies such as satellite surveillance systems or maritime drones can revolutionise maritime border surveillance and the fight against piracy.

Economic, environmental and geopolitical changes may require an adaptation of maritime practices, such as the adjustment of maritime transport processes in the face of the rise of e-commerce. A break in methods may also be motivated by the recognition of inefficient or unsustainable current practices, such as overfishing, leading to stricter regulations. These changes often require a change in mentality and the political will to adapt, involving investment in research, regulatory reform and collaboration between players in the maritime sector.

When General de Gaulle decided in 1966 to withdraw France from NATO's integrated command, he demonstrated a break with the past. This decision, motivated by the search for strategic autonomy and independence from the American ally, is illustrative of what François Géré[54] refers to as a State's political autonomy. In his view, political autonomy is compromised when a state does not possess nuclear weapons and has to rely on those who do for its security. Ambitious in its scope, this approach could be a benchmark for a country like Benin in its quest for full sovereignty.

For an effective maritime strategy, it would be both innovative and promising to start by developing a naval doctrine. This doctrine could focus on protecting Benin's maritime territory by anticipating and preventing potential threats. In practical terms, this would mean acting proactively to prevent the emergence of these threats in Benin's waters. This ambitious approach could be broken down into two main areas:

- The first stage, detection, involves strengthening the French Navy's surveillance and detection resources, such as semaphores, drones and aircraft, to spot any suspicious activity beyond the waters under national sovereignty;
- Once a suspicious activity has been detected, the next step is to take action. This requires appropriate means of intervention to respond effectively, both at sea and on land, including administrative and jurisdictional measures. It is essential to identify and know the structures and players involved in this chain to ensure an appropriate response.

This approach to the sanctuary of national territorial waters aims to create a safe maritime environment, free from criminal threats, thanks to a multifaceted detection and intervention system. Although some threats may remain, particularly those from supposedly safe vessels already present in territorial waters, this system should enable an effective response.

[54] François Géré, La pensée stratégique française contemporaine, Paris Economica, 9 February 2017, p. 20.

In addition, to follow the logic of a break in the way things are done, it is recommended that a maritime research centre or institute be set up. In addition to its study and advisory functions, this centre would serve as a doctrinal reference in maritime matters. Given that Benin is a country with a maritime vocation, it is imperative to transform this vocation into reality to ensure its socio-economic development.

Adopting such an approach marks a break with our usual practices and reflects our strong will, a key element in our ability to act. It gives us autonomy vis-à-vis our technical and financial partners, while at the same time confronting us with the need to make up for the current lack of resources in the French Navy, the linchpin of maritime safety in Benin. This change in our methods, by following the steps described above, will enable us to project ourselves into the distant future.

1 .2: Ability to project far into the future.

The ability to project far into the future is essential for maritime strategies, as it implies the ability to anticipate and plan beyond immediate challenges and opportunities. This requires long-term vision, forward-looking analysis and strategic decision-making. Indeed, the Benin Alafia 2025 Forward View, drawn up in 2000, illustrates a long-term projection for the country. This document details the aspirations of the Beninese people for the next twenty-five years. As it comes to an end, a similar exercise is underway for 2060, underlining the importance of long-term planning for Benin's development. The foresight initiative aims to strengthen national capacities for anticipation and action in the face of current and future changes, both nationally and internationally.

The horizon chosen for the forward-looking vision is the year 2060, marking one hundred years since Benin's independence, in order to project the country's desired development. The Minister of State for Development and the Coordination of Government Action, Abdoulaye Bio Tchané[55] , stresses that this new vision should make it possible to devise a long-term national development strategy to meet the major challenges facing Beninese society, including the promotion of peace, poverty reduction, the fight against inequality, environmental preservation, and many others. He urges us to dream big for Benin and to include this vision in all ministerial spheres.

[55] Abdoulaye Bio Tchané, Minister of State, in charge of Development and Coordination of Government Action, Vice-Chairman of the National Committee for Orientation and Supervision of Forward Thinking, Palais des Congrès, Cotonou, 23 November 2023.

Updating the SNPSSM towards a blue economy requires a long-term vision. National stakeholders must see this process as a sustainable development challenge, ensuring that maritime resources are exploited while preserving the safety and security of Benin's maritime sector. It is crucial to project the maritime sector far into the future in order to guarantee sufficient resources for future generations.

To achieve this, it is necessary to have a clear idea of what might be termed Benin's "vital interests". The sea and all related activities are of vital interest to a country like Benin, whose economy is closely linked to the exploitation of the sea's resources and opportunities. While it is "*the responsibility of the Head of State to constantly assess the limits of our vital interests*"[56] , it is also clear that to enable him to make this assessment, he must be able to rely on key elements, including the country's capacity to project itself far into the future

The new blue economy strategy currently being drawn up must therefore become a genuine planning instrument capable of projecting the maritime sector into the distant future. The year 2060, for example, could be chosen as the horizon to be in phase with the "2060 vision" adopted at national level.

The quest for coherence in Benin's maritime strategy highlights the importance of maintaining regional and international cooperation in this area. Given that the sea is a major geopolitical challenge and that maritime safety issues require collaborative approaches, Benin must remain committed to this cooperation. What's more, given the country's current capacity to respond to maritime insecurity, it is vital to maintain this cooperation.

Section 2: Maintaining regional and international cooperation.

Because the sea is a common heritage of mankind, because the threats to the sea, its resources and the countries bordering the Gulf of Guinea are transboundary, and because the contemporary world is based on public international law, Benin acts in the maritime field in compliance with the international rules laid down in this area and is bound by a number of regional and international cooperation instruments and mechanisms.

The implementation of these various cooperation instruments requires cooperative practices on the part of the States involved. For example, the SOLAS (Safety of the Life at Sea)

[56] Statement by President Jacques Chirac on 19 January 2006 in Brest on nuclear deterrence, consulted on www.vie-publique.fr on 15 April 2024.

Convention is designed to protect human life at sea. It came into being following the sinking of the TITANIC off the coast of Newfoundland (Canada) after colliding with an iceberg, which killed around 1,500 of the 2,200 people on board. It defines standards for construction, equipment, safety and security management (ship and ports), and special measures for special-purpose ships.

Moreover, in the specific case of the countries of the Gulf of Guinea, there is the ECOWAS Integrated Maritime Strategy (IMS), which emphasises a people-centred response to the management and exploitation of the maritime domain.

Operating in an interdependent environment, maintaining regional and international cooperation will enable Benin to gain experience to improve its maritime strategy.

2.1: An interdependent environment.

The world order inherited from the Treaties of Westphalia in 1648 makes the State the central player on the international stage. In this system, which operates on the basis of conventions, treaties, agreements and memorandums of understanding, certain areas, such as the sea, are entrusted to specialised agencies of the United Nations Organisation, which was set up in the aftermath of the Second World War to promote international peace and security. The International Maritime Organisation (IMO), created by a convention that came into force in 1958, is the specialised agency of the United Nations responsible for ensuring the safety and security of maritime transport and preventing pollution of the marine environment by ships.

Benin, which has a coastline and is a full member of these conventions, is aware that it is part of an interdependent maritime environment. Indeed, interdependence is not only a consequence of the globalised economy, but also a necessity for States to self-regulate their quest for power. In this changing world, where the quest for independence or what is known as "sovereignism" is increasingly asserted, particularly in countries south of the Sahara, "*thinking about independence means thinking about interdependence*"[57] i.e. thinking about our relationship with others as part of our desire for independence. Independence cannot be hermetic. In fact, the sea is the domain par excellence of interdependence.

In his strategic breviary, Hervé Couteau-Bégarie states that "*the master of the sea, dependent on his maritime trade, needs to keep his lines of communication open*".[58] It is therefore impossible to be self-sufficient in the maritime environment. Mutual dependence in

---

[57] Bonaventure Mvé Ondo, "*Dépendance, indépendance, interdépendance, repenser l'indépendance*", Afrique contemporaine n°271-272, 2002, pp.19-31, accessed at www.cairn.info on 13 April 2024.
[58] Hervé Couteau-Bégarie, op.cit, p.111.

this sector is not only a necessity, but also a condition for achieving national objectives. Moreover, in the absence of a defence industry, interdependence is even greater.

Taking into account the interdependent environment in which the national maritime approach evolves should strengthen our determination to equip ourselves with a credible capability. In addition, the opportunities offered by continued regional and international cooperation will enable the national maritime safety system to gain experience.

2 .2: Acquiring experience.

The African state in its modern sense is an import of the state that emerged from the Treaties of Westphalia in 1648. Bertrand Badie calls it an "imported state".[59] This state, which is not the result of African dynamics, is the one that will have to govern the life of newly independent countries. In this context, it is easy to understand that everything has to be learned. Learning how to implement the State, i.e. how to exercise its prerogatives according to the standards defined in the West, is the fundamental challenge facing African societies. This is all the more true given that, at the time of independence, the prevailing state order was not disrupted. At best, there was a change of rulers: the colonialists were replaced by the natives. Under these conditions, the know-how and technologies needed to ensure maritime safety, for example, come from the West, which provides standards and equipment.

Maintaining regional and international cooperation in the maritime sector is therefore a necessity for the French Navy in order to gain experience. In concrete terms, since Benin has no naval school and no industrial unit capable of manufacturing the slightest nautical vector, it can only maintain a naval force by relying on the know-how of others. Better still, to gain experience in the sector, it needs to develop partnerships with countries that have a credible naval capability.

The recent organisation of Operation Safe Domain II can be analysed from this angle. While it is true that this joint patrol aims first and foremost to assert the presence of the naval forces of the countries in zone E in their common maritime space in the context of maritime security, it is also true that the active participation of the French Navy in this exercise enables it to gain experience alongside the other navies involved. In fact, in addition to the interoperability of the equipment and procedures tested during this exercise, the human resources involved also gained experience in terms of discovering equipment that is not

[59] Bertrand Badie, *"l'État importé, l'occidentalisation de l'ordre politique"*, Fayard, Paris, 1992, 334p.

available, the naval aviation dimension of such an operation and above all in terms of "drill".[60] This can be said of the French Navy's participation in all other initiatives of this kind.

Photo N°8 : Beninese fusiliers commandos exercise with French elements

Source: French Navy Operations Division,2021

The recent joint exercise between Beninese fusiliers commandos and French elements illustrates the importance of international military cooperation and the exchange of skills between armed forces. This joint training aims to strengthen the operational capabilities of Benin's fusiliers commandos by giving them the opportunity to benefit from the expertise and advanced techniques of the French military.

Partial conclusion

The implementation of Benin's maritime strategy faces major challenges and promising prospects. A study of the security arrangements in the Gulf of Guinea highlights these challenges and the opportunities to be seized.

Firstly, it is essential to recognise the structural shortcomings at national level, such as the lack of coordination and resources, which hamper the effectiveness of maritime operations. These challenges require integrated solutions, involving the strengthening of national capabilities, in particular through investment in adequate equipment and the establishment of a marine research centre.

Regional dynamics add further complexity, with potential tensions between riparian countries and cooperation challenges. Yet regional and international cooperation remains a

---

[60] Military slang for an ability acquired through repeated training. Someone is said to be drilled in a skill when he or she executes or implements that skill with ease. Drill is the only way to ensure ease of execution.

crucial prospect for maritime security and sustainable development in the Gulf of Guinea. It is imperative to maintain and strengthen these partnerships, while ensuring that Benin's national interests are preserved.

Finally, implementing Benin's maritime strategy requires a holistic approach, integrating both national measures and regional and international cooperation efforts. By overcoming the challenges identified and seizing the opportunities ahead, Benin can promote effective and sustainable management of its maritime resources, thereby contributing to the security and prosperity of the region as a whole.

GENERAL CONCLUSION

The study of Benin's security arrangements in the Gulf of Guinea highlights a series of complex and interconnected challenges. Structural shortcomings at national level, including a lack of coordination between the various entities and limited resources, hamper the effectiveness of maritime operations. At the same time, regional dynamics exacerbate the situation, with potential tensions between riparian countries and cooperation that is sometimes difficult to implement. To overcome these challenges, a number of actions are required. Firstly, it is imperative to strengthen national capabilities, in particular by investing in adequate equipment for the French Navy and establishing a marine research centre.

Regional and international cooperation must be maintained and strengthened, while ensuring that Benin's national interests are preserved. The country's maritime approach needs to be rethought, taking into account current challenges and looking to the future through a long-term vision. In 2006, during his term as President of France, Jacques Chirac stressed the importance of not reducing the complexity of defence and security issues solely to the fight against terrorism.[61] He warned against the temptation to focus all attention on this specific threat to the detriment of other issues. Chirac stressed that even in the face of the emergence of new threats, these should not overshadow other existing challenges.

This reflection highlights the need not to underestimate maritime insecurity in Benin, despite the urgency of the land-based security challenge on the country's northern borders. It stresses that, although the land challenge is currently of greater concern, maritime insecurity should not be ignored. The aim of this analysis is to examine Benin's security system in the face of maritime insecurity in the Gulf of Guinea, identify its structural weaknesses and propose improvements to make it more effective.

After examining the national security system, it emerges that it is divided into two main dimensions: institutional and operational. The institutional dimension is based on the national maritime protection, safety and security strategy, centred around the National Authority in charge of State Action at Sea via the Maritime Prefecture. The operational dimension of the system relies mainly on the French Navy. State actions at sea, initiated by the Préfecture Maritime, are carried out by units of the French Navy, and conversely, the French Navy can also initiate actions falling within the remit of the Préfecture Maritime. However, the operation of this system has highlighted a number of shortcomings, in particular the

[61] Statement by President Jacques Chirac on 19 January 2006 in Brest on nuclear deterrence, consulted on www.vie-publique.fr on 15 April 2024.

ineffectiveness of the inter-administration coordination required by the National Authority in charge of State Action at Sea, the inadequacy of the operational resources of the French Navy and limited financial and material resources.

In addition to the internal challenges identified, it is worth considering the complexity of the existing regional institutional mechanisms. While the regional security approach is laudable, its day-to-day implementation can sometimes lack clarity. In Zone E, for example, the individual ambitions of each member country can hamper cooperation. Nigeria, aware of its naval superiority over neighbours such as Benin and Togo, has had border disputes with Cameroon over the Bakassi Peninsula. Benin, whose oil reserves are entirely offshore[62] and shared with Nigeria, therefore has an interest in understanding Nigeria's perspectives on the Law of the Sea and related issues.

As a country with a maritime vocation, Benin cannot ignore the importance of developing a naval doctrine. This maritime dimension, which is not yet fully integrated into the current maritime strategy, requires in-depth consideration by the various stakeholders. To move towards a blue economy, it is imperative to review existing approaches and recognise our vital interests, among which the sea and its activities play a crucial role. This requires a proactive decision to acquire adequate equipment to enhance the effectiveness of the French Navy. At the same time, we need to create a marine research centre or institute. In addition to its research and advisory functions, it would become the benchmark for the country's naval doctrine.

In the current context, the coherence of Benin's maritime strategy requires the maintenance of existing regional and international cooperation. Cooperation is not an obstacle in itself, but it is the national structural weakness that makes the country too dependent on this cooperation. It is crucial to strengthen national capacities so that we can choose the forms of cooperation that best suit our interests. Cooperation often prioritises the interests of technical and financial partners over those of the beneficiary country, a dynamic that needs to be corrected.

In their relationship with the sea, Beninese people often display inappropriate attitudes, ranging from deliberate acts of pollution to illegal fishing practices and even the uncontrolled installation of leisure infrastructures near the water. This often uncivil attitude is largely due to a lack of education about the sea. This is all the more regrettable given that the country's geography and economy are largely dependent on its maritime environment. To ensure sustainable development, education about the sea from an early age is essential. Updating the

[62] Philippe Noudjenoume, op. cit.

National Strategy for Maritime Protection, Safety and Security should provide an opportunity to rethink Benin's approach to the sea, in line with a long-term vision such as "Vision 2060".

A holistic and multidimensional approach is fundamental to addressing security challenges and promoting sustainable development in the Gulf of Guinea. By combining national and international efforts, better management of maritime resources can be ensured, contributing to the security and prosperity of the region as a whole.

BIBLIOGRAPHY

A. GENERAL WORKS

1- Badie Bertrand, *L'État importé, occidentalisation de l'ordre politique*, Paris, Fayard, 1992, 334 p.

2- Couteau-Bégarie Hervé, *Bréviaire stratégique*, Paris, Editions du Rocher, 2016, 134p.

3- Géré François, *La pensée stratégique française contemporaine*, Paris Economica, 300p.

4- General Sir Rupert Smith, *L'utilité de la force, l'art de la guerre aujourd'hui*, Paris Economica, 2007, 212p.

5- KAMTO Maurice, *L'urgence de la pensée, réflexions sur une précondition du développement en Afrique*, Yaoundé, éditions MANDARA, 1993, 210p.

6- Taillat Stéphane - Henrotin Joseph- Schmitt Olivier (coll), *Guerre et stratégie*, France, PUF, 2015, p.287.

7- TSHIYEMBE Mwayila and BUSAKA Mayele, *l'Afrique face à ses problèmes de sécurité et de défense,* Paris, Présence africaine, 2001, 262p.

B. SPECIALIST LITERATURE

8- Amegée, K., *Sécurité maritime dans le golfe de Guinée : une réponse régionale à un défi mondial*, Revue des études africaines contemporaines, 2020.

9- International Maritime Bureau, *Piracy and armed robbery against ships: Annual Report 2020.*

10- COUTEAU-BEGARIE Hervé, *Le meilleur des ambassadeurs, théorie et pratique de la diplomatie navale*, Paris, Economica, 2010, 384p.

11- HENROTIN Joseph, *Les fondements de la stratégie navale au XXIe siècle*, Paris, Economica, 2011 488 p.

12- White Paper on Defence and Security, France, 2013 edition, 160 p.

13- Ministère de la Défense Nationale, *Actes finaux des états généraux de la Défense*, Cotonou, 14 - 18 July 1997, 146p.

14- Ministère de la Défense Nationale, *Forum de réflexion géostratégique, Les actes du Forum*, Cotonou, 2003, 323p.

15- NOUDJENOUME Philippe (dir), *Les frontières maritimes du Bénin*, Paris, l'Harmattan, 2004, 151p.

16- Onuoha, Freedom C, *Maritime security in the Gulf of Guinea: a complex 'systemic security' challenge*, African Security Review, 28(3), 2019, pp.238-252.

17- United Nations Office on Drugs and Crime, *Global Migrant Smuggling Study 2018.* UNODC, 2019.

18- Oluwaniyi, O. O, *Oil theft and maritime security in the Gulf of Guinea: A review of Nigeria's efforts to combat the scourge*, African Security Review, 26(1), 2017, pp.80-95.

C. REPORTS, STUDIES, ARTICLES, REVIEWS AND COMMUNICATIONS

19- ABDELHAK Bassou, *La mer du golfe de Guinée, richesses, conflits et insécurité*, Revue Maroco-espagnole de droit international et relations internationales N°2, 2014, pp.151-163.

20- Alexandra Bellayer Roille, *Les enjeux politiques autour des frontières maritimes*, CERISCOPE Frontières, 2011, 18p, consulted on 23/03/2024.

21- Anselme Tsassa, Gulf of Guinea: *political limits and geopolitical stakes?* Analysis note, Thinking Africa n°32, October 2015.

22- Dr Charles UKEJE, Pr Wullson MVOMO ELA, *Approche africaine de la sécurité maritime : cas du golfe de Guinée*, Abuja, Fondation Fridriech Ebert STIFTUNG, 2013, 50p.

23- Ilinca Mathieu, *Le CIC, clé de voûte de l'architecture interrégionale de sécurité dans le golfe de Guinée*, Revue Défense nationale n°792, Editions Comité d'études de défense nationale, 2016, pp.93-98.

24- Jean Dominique Giuliani, *l'Europe a-t-elle une stratégie maritime*, Revue Défense nationale n°789, Editions Comité d'études de défense nationale, 2016/4, pp.31-36.

25- Hervé Moulinier, *La stratégie maritime de la France et ses perspectives*, Annales des mines - Responsabilités et environnement n°70, éditions ESKA, 2013/2, pp.81-87.

26- Hugues Eudeline, *Stratégie maritime : évolutions et nouveaux acteurs*, Revue Défense nationale n°789, Editions Comité d'études de défense nationale, 2016/4, pp.37-43.

27- Michel Luntumbue, *Piraterie et insécurité dans le golfe de Guinée : défis et enjeux d'une gouvernance maritime régionale*, GRIP analysis note, Brussels, 30 September 2011, www.grip.org accessed on 23 March 2024.

28- Serge Ségura, *Une stratégie maritime internationale: mythe ou réalité*, Revue Défense nationale n°789, Editions Comité d'études de défense nationale, 2016/4, pp.24-30.

29- Thierry Vircoulon, *Violette Tournier, Sécurité dans le golfe de Guinée : un combat régional*, Politique étrangère 2015/3, pp.161-174.

D. CONSTITUTIONAL, LEGISLATIVE AND REGULATORY TEXTS

30- Act no. 90-32 of 11 December 1990 establishing the Constitution of the Republic of Benin, as revised by Act no. 2019-40 of 07 November 2019.

31- Act no. 2020-15 of 03 July 2020 amending and supplementing Act no. 90-016 of 18 June 1990 creating the Beninese Armed Forces.

32- Decree no. 2021-579 of 03 November 2021 on the general organisation of the Beninese Armed Forces and the organisation of command in the FAB.

33- Decree no. 2008-735 of 22 December 2008 approving the national security policy and strategy.

34- Decree no. 2014-785 of 31 December 2014 on the creation, powers and operation of the national authority responsible for State action at sea, as amended by Decree no. 2017-523 of 15 November 2017 and Decree no. 2019-450 of 09 October 2019.

35- Decree no. 2007-621 of 31 December 2007 on the creation, composition, powers, organisation and operation of the Maritime Protection, Safety and Security Management Bodies.

36- Decree no. 2021-581 of 03 November 2021 on the powers, organisation and operation of the National Navy Headquarters.

37- Decree no. 2013-551 of 30 December 2013 adopting the National Strategy for Maritime Protection, Safety and Security.

38- Decree n°2020-270 of 06 May 2020 on the obligation of an armed guard for commercial vessels bound for Benin ports.

39- Interministerial order no. 2020-016 on the protection of ships in Benin's territorial waters.

E. WEBOGRAPHY - QUESTIONNAIRE

40- www.gouv.bj.

41- www.vie-publique.fr/discours/160098-declaration-de-m-jacques-chirac-president-de-la-republique-sur-la-pol
42- WWW.cairn.info
43- www.news.un.org: Transnational maritime crime is growing in scale and sophistication, accessed May 2024.
44- www.atlas-mag.net, Atlas Magazine: Maritime piracy: causes, issues and instruments to combat the phenomenon, consulted in April 2024.
45- www.atlas-mag.net, Atlas Magazine: Nigeria's waters: the new global centre of maritime piracy, consulted in April 2024.
46- www.gouvernement.fr: General Secretariat for the Sea, State action at sea, consulted

Appendices

Appendix No.1 : Questionnaire

Part 1: Examination of the response mechanism

1. What types of stakeholders are involved in maritime safety in Benin?
   ☐ Government ☐ Port authorities
   ☐ Navy ☐ Private companies
   ☐ NGOS ☐ Others : __________

2. How would you rate the speed of the response in the event of a maritime incident?
   ☐ Very quick ☐ Fairly fast ☐ Average
   ☐ Slow ☒ Very slow

3. Are the resources allocated to maritime safety sufficient
   ☐ Yes ☐ No
   If no, what resources are you mainly lacking?
   ☐ Human ☐ Material ☐ Financial

4. Is the maritime safety training offered to stakeholders adequate?
   ☐ Yes ☐ No

Part 2: Identification of weaknesses in the response system

5. What is the main obstacle to an effective maritime safety response?
   ☐ Lack of coordination between players ☐ Insufficient resources
   ☐ Inadequate legislative framework ☐ Lack of training

6. To what extent do technological challenges affect response capacity?
   ☐ Very significantly ☐ Significantly
   ☐ Moderately ☐ Slightly ☐ Not at all

7. Does the current legislation effectively support maritime safety?
   ☐ Yes, very effectively ☐ Yes, fairly effectively
   ☐ Moderately ☐ No, not very effectively
   ☐ No, not at all effective

Part 3: Proposed solutions to improve the response system

8. What improvements would you prioritise to enhance maritime safety?
   ☐ Technological improvements
   ☐ Increase in financial resources
   ☐ Strengthening legislation

☐ Increased training and awareness
☐ Improved coordination between stakeholders

9. Should regional maritime safety initiatives be strengthened?
☐ Yes, strongly ☐ Yes, moderately ☐ Neutral
☐ No, not much ☐ No, not at all

10. How effective are international cooperation mechanisms perceived to be in managing maritime safety?
☐ Very effective ☐ Somewhat effective ☐ Moderately effective
☐ Not very effective ☐ Not at all effective

Appendix No.2 : Decree No. 2014-785 of 31 December 2014.

APJB/

REPUBLIQUE DU BENIN

Fraternité-Justice-Travail

PRESIDENCE DE LA REPUBLIQUE

DECRET N°2014-785 DU 31 DECEMBRE 2014 portant création, organisation, attributions et fonctionnement de l'Autorité Nationale Chargée de l'Action de l'Etat en Mer.

LE PRESIDENT DE LA REPUBLIQUE,

CHEF DE L'ETAT,

CHEF DU GOUVERNEMENT,

Vu la loi n° 90-32 du 11 décembre 1990 portant Constitution de la République du Bénin ;

Vu la Convention des Nations-Unies sur le droit de la mer, signée à Montégo Bay le 10 décembre 1982 ;

Vu la loi n° 97-028 du 15 janvier 1999 portant organisation de l'administration territoriale de la République du Bénin ;

Vu la loi n° 2010-11 du 27 décembre 2010 portant code maritime en République du Bénin ;

Vu la proclamation, le 29 mars 2011 par la Cour Constitutionnelle, des résultats définitifs de l'élection présidentielle du 13 mars 2011 ;

Vu le décret n°2014-512 du 20 août 2014 portant composition du Gouvernement ;

Vu le décret n° 2012-191 du 03 juillet 2012 fixant la structure-type des Ministères ;

Vu le décret n° 2014-260 du 18 avril 2014 portant attributions, organisation et fonctionnement du Ministère de la Défense Nationale ;

Vu le décret n° 2012-432 du 06 novembre 2012 portant attributions, organisation et fonctionnement du Ministère Délégué auprès du Président de la République Chargé de l'Economie Maritime, du Transport Maritime et des Infrastructures Portuaires ;

Vu le décret n° 2013-551 du 30 décembre 2013 portant adoption de la stratégie nationale de protection, de sécurité et de sûreté maritimes ;

Sur proposition conjointe du Ministre de la Défense Nationale et du Ministre de l'Economie Maritime et des Infrastructures Portuaires ;

Le Conseil des Ministres entendu en sa séance du 25 novembre 2014,

D E C R E T E :

Chapitre 1er : DE LA CREATION, DE LA MISSION ET DES ATTRIBUTIONS.

Article 1er : Il est créé à la Présidence de la République une structure dénommée « Autorité Nationale Chargée de l'Action de l'Etat en Mer » (ANCAEM).

Article 2 : L'Autorité Nationale Chargée de l'Action de l'Etat en Mer (ANCAEM) est dotée de la personnalité juridique, de l'autonomie financière et de l'autonomie de gestion.

1

L'autorité du Préfet Maritime s'exerce jusqu'à la limite des eaux sur le rivage de la mer. Elle ne s'exerce pas à l'intérieur des limites administratives des ports. Dans les estuaires, elle s'exerce en aval des limites transversales de la mer.

Article 8 : Le Préfet Maritime est nommé par le Président de la République parmi les officiers généraux des Forces Navales.

Il a rang de ministre. Il dispose d'un cabinet.

Section 2 : Du Secrétaire Général

Article 9 : Le Secrétaire Général est la deuxième personnalité de l'Autorité Nationale Chargée de l'Action de l'Etat en Mer (ANCAEM). Il supplée le Préfet Maritime en cas d'absence ou d'empêchement.

Article 10 : Le Secrétaire Général constitue la mémoire de l'Autorité Nationale Chargée de l'Action de l'Etat en Mer (ANCAEM). Il assure le fonctionnement harmonieux des Services Techniques de l'institution.

Il est nommé par le Président de la République parmi les cadres civils ayant des connaissances avérées dans le domaine maritime.

Section 3 : Du pôle des experts

Article 11 : Le pôle des experts est chargé d'assister le PREMAR dans ses missions de gouvernance, de coordination et de police administrative en mer. Il conseille le PREMAR sur l'action de l'Etat en mer et prépare les réunions du Comité Technique de Protection, de Sécurité et de Sûreté Maritime.

Article 12 : Il est composé d'une équipe constituée d'experts issus des principales administrations engagées dans l'Action de l'Etat en Mer à savoir :

- la Direction de la Marine Marchande ;
- la Police Nationale ;
- la Protection civile ;
- la Gendarmerie Nationale ;
- les Douanes ;
- les Forces Aériennes ;
- les Forces Navales ;
- la Direction des Pêches ;
- la Direction Générale de l'Environnement ;
- le Secrétariat Général du Ministère des Affaires Etrangères, de l'Intégration Africaine, de la Francophonie et des Béninois de l'Extérieur ;
- le Secrétariat Général du Ministère de la Justice, de la Législation et des Droits de l'Homme.

Section 4 : Des services techniques

Article 13 : Les services techniques de l'ANCAEM sont :

- le service de la Sécurité et de la Sûreté Maritimes ;
- le service de l'Information et de la Communication Maritime ;
- le service Juridique ;
- le service de l'Environnement Marin et de la Recherche Océanographique ;
- le service des Affaires Administratives et Financières ;

- le Secrétariat Administratif.

Article 14 : Le service de la sécurité et de la sûreté maritime assure la sauvegarde des droits souverains du Bénin dans l'espace maritime conformément au droit national et international.

Article 15 : Le service de l'Information et de la Communication Maritime est chargé de promouvoir l'image de marque de l'ANCAEM et de l'espace maritime béninois auprès de la communauté maritime nationale et internationale.

Article 16 : Le service juridique est chargé de conseiller le Préfet Maritime sur le cadre légal dans lequel s'exercent l'ensemble de ses prérogatives.

Article 17 : Le Service de l'Environnement Marin et de la Recherche Océanographique est chargé de la protection de la biodiversité marine et d'assurer la connaissance et la maitrise du milieu marin, à travers des recherches et des publications, en liaison avec les universitaires et les centres universitaires nationaux et étrangers.

Article 18 : Le Service des Affaires Administratives et Financières est chargé d'assurer la gestion des ressources humaines, financières et du matériel de l'Autorité Nationale Chargée de l'Action de l'Etat en Mer (ANCAEM). Il dispose d'un Agent comptable.

Article 19 : Le Secrétariat administratif est chargé de toutes les activités d'ordre administratif indispensables au bon fonctionnement de l'Autorité Nationale Chargée de l'Action de l'Etat en Mer (ANCAEM). Il s'occupe de l'enregistrement des courriers et leur ventilation, de la saisie des documents, de l'accueil et de l'orientation des usagers.

Chapitre 3 : DU COMITE TECHNIQUE DE PROTECTION DE SECURITE ET DE SURETE MARITIMES (CTPSSM)

Article 20 : Le Comité Technique de Protection, de Sécurité et de Sûreté Maritimes (CTPSSM) assiste l'ANCAEM dans sa mission de coordination des administrations. Il est présidé par le PREMAR.

Article 21 : Le Comité technique de protection, de sécurité et de sûreté maritimes comprend, outre son président :

- ✓ un coordonnateur/Secrétaire permanent : le Chef d'Etat-Major des Forces Navales ;
- ✓ un coordonnateur adjoint : le Directeur de la Marine Marchande ;
- ✓ un rapporteur : le Directeur Général du Port Autonome de Cotonou ;
- ✓ des membres à savoir :
  - le Directeur des Transports Maritimes et fluvio-lagunaires ;
  - le Directeur Général des Douanes et Droits Indirects ;
  - le Directeur des Pêches ;
  - le Directeur Général de l'Environnement ;
  - le Directeur de l'Agence Nationale de la Prévention et de la Protection Civile ;
  - le Directeur Général de l'Agence Nationale de Gestion Intégrée des Espaces Frontaliers ;

- le Directeur des Affaires Juridiques au Ministère des Affaires Etrangères, de l'Intégration Africaine, de la Francophonie et des Béninois de l'Extérieur ;
- le Directeur des Pays du Voisinage au Ministère des Affaires Etrangères, de l'Intégration Africaine, de la Francophonie et des Béninois de l'Extérieur ;
- le Directeur des affaires civiles et Pénale du Ministère de la justice ;
- le Chef d'Etat-Major des Forces Aériennes ;
- le Directeur Général du Budget ;
- le Commissaire chargé du Commissariat spécial de police du port de Cotonou, représentant le Directeur Général de la Police Nationale ;
- le commandant de la brigade spéciale de la gendarmerie du port de Cotonou, représentant le Directeur Général de la Gendarmerie Nationale.

Article 22 : Le Comité Technique de Protection, de Sécurité et de Sûreté Maritimes se réunit une fois par trimestre en session ordinaire, sur convocation de son Coordonnateur/Secrétaire permanent. Il peut également se réunir en session extraordinaire en cas de besoin.

Le coordonnateur est suppléé dans sa mission par son adjoint en cas d'absence ou d'empêchement.

Article 23 : Sous la responsabilité de l'ANCAEM, les Forces Navales du Bénin peuvent intervenir en mer au profit des autres administrations de l'Etat. Les programmes et les modalités de ces actions sont établis par le CTPSSM. Les Forces Navales peuvent embarquer pour ces missions des agents de ces administrations.

Chapitre 4 : DU PERSONNEL ET DU BUDGET DE L'AUTORITE NATIONALE CHARGEE DE L'ACTION DE L'ETAT EN MER (ANCAEM).

Section 1 : Du personnel

Article 24 : Le personnel de l'Autorité Nationale Chargée de l'Action de l'Etat en Mer (ANCAEM) est mis à sa disposition par les Ministres concernés par l'Action de l'Etat en Mer.

Il comprend :

- les Agents Permanents de l'Etat, civils, militaires et paramilitaires ;
- les agents contractuels de l'Etat ;
- les agents conventionnés.

Article 25 : Les modalités de recrutement du personnel et les qualifications exigées sont définies par le Préfet Maritime dans le cadre de la politique générale et conformément aux textes en vigueur.

Section 2 : Du budget

Article 26 : Les ressources de l'Autorité Nationale Chargée de l'Action de l'Etat en Mer (ANCAEM) comprennent :

- des ressources du budget national ;
- des ressources du mécanisme de financement de la sécurisation maritime ;
- des ressources provenant des activités telles que les arraisonnements sur les fautes de police en mer et la gestion des catastrophes civilement imputables ;
- des subventions, dons et legs, conformément à la législation en vigueur ;

- une partie du produit des amendes, transaction et confiscations prononcées pour la répression des infractions commises en mer.

Article 27 : le règlement financier de l'Autorité Nationale Chargée de l'Action de l'Etat en Mer (ANCAEM), approuvé par le Ministre chargé des Finances, détermine la portée de l'autonomie financière et de gestion de l'Autorité Nationale Chargée de l'Action de l'Etat en Mer (ANCAEM) notamment en ce qui concerne :

- les règles de préparation et de présentation du budget de l'Autorité Nationale Chargée de l'Action de l'Etat en Mer (ANCAEM) qui devra être approuvé par le Ministre chargé des Finances ;
- les opérations ou actes de gestion soumis au visa préalable du Ministre chargé des Finances ;
- les modalités de contrôle du Ministre chargé des Finances à l'égard du gestionnaire administratif et financier ainsi que de l'agent comptable.

Article 28 : Les charges de l'Autorité Nationale Chargée de l'Action de l'Etat en Mer (ANCAEM) comprennent :

- les dépenses de fonctionnement ;
- les dépenses d'investissement.

Article 29 : L'Autorité Nationale Chargée de l'Action de l'Etat en Mer (ANCAEM) est soumise au contrôle des organes de contrôle et notamment de l'Inspection Générale d'Etat.

Chapitre 5 : **DES DISPOSITIONS DIVERSES ET FINALES.**

Article 30 : Le Préfet Maritime, le Secrétaire Général et les Chefs des Services Techniques bénéficient d'avantages matériels et financiers qui seront définis par arrêté.

Article 31 : Les membres du CTPSSM perçoivent une indemnité de session à l'occasion de leurs réunions périodiques. Le montant de cette indemnité est déterminé par le règlement financier.

Article 32 : Le Ministre de la Défense Nationale, le Ministre de l'Intérieur, de la Sécurité Publique et des Cultes, le Ministre de l'Economie, des Finances et des Programmes de Dénationalisation et le Ministre de l'Economie Maritime et des Infrastructures Portuaires sont chargés, chacun en ce qui le concerne, de l'exécution du présent décret qui sera publié au Journal Officiel de la République du Bénin.

Fait à Cotonou, le 31 decembre 2014

Le Président de la République,
Chef de l'Etat, Chef du Gouvernement,

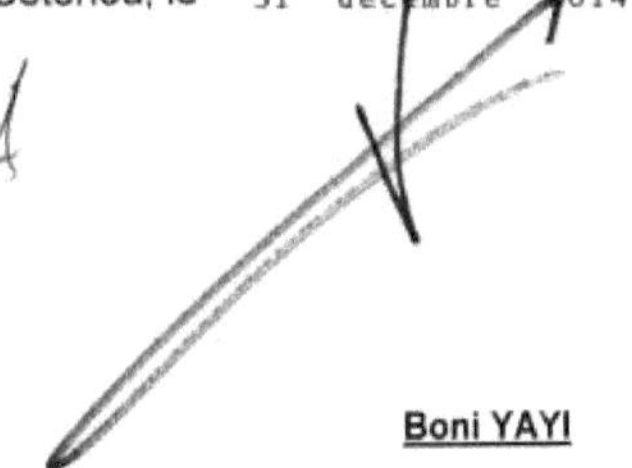

Boni YAYI

Le Ministre d'Etat chargé de l'Enseignement Supérieur
et de la Recherche Scientifique,

François Adebayo ABIOLA

Le Ministre de l'Economie, des Finances
et des Programmes de Dénationalisation,

Komi KOUTCHE

Le Ministre de la Défense Nationale,

Robert Théophile YAROU

Le Ministre des Affaires Etrangères,
de l'Intégration Africaine, de la Francophonie
et des Béninois de l'Extérieur,

Nassirou BAKO ARIFARI

Le Ministre de l'Intérieur, de la
Sécurité Publique et des Cultes,

Dossou Simplice CODJO

Le Ministre de l'Economie Maritime
et des Infrastructures Portuaires,

Rufin Orou Nan NANSOUNON

AMPLIATIONS : PR 6 AN 4 CS 2 CC 2 CES 2 HAAC 2 HCJ 2 MECESRS 2 MDN 2 MAEIAFBE 2 MEFPD2 MISPC 2 MEMIP 2 AUTRES MINISTERES 21 SGG 4 DGBM-DCF-DGTCP-DGID-DGDDI 5 BN-DAN-DLC 3 GCONB-DGCST-INSAE 3 BCP-CSM-IGAA-IGE 4 UAC-ENAM-FADESP 3 UP-FDSP2 JORB 1

Appendix No.3 : Interministerial Order n°2020016/MIT/MDN/MISP/MEF/DC/SGMCTJ/SA /020SGG20 of 13 July 2020.

RÉPUBLIQUE DU BÉNIN

MINISTÈRE DES INFRASTRUCTURES ET DES TRANSPORTS

MINISTÈRE DÉLÉGUÉ AUPRÈS DE LA PRÉSIDENCE DE LA RÉPUBLIQUE, CHARGÉ DE LA DÉFENSE NATIONALE

MINISTÈRE DE L'INTÉRIEUR ET DE LA SÉCURITÉ PUBLIQUE

MINISTÈRE DE L'ÉCONOMIE ET DES FINANCES

ARRÊTÉ INTERMINISTÉRIEL

ANNEE 2020 N° 016 /MIT/MDN/MISP/MEF/DC/SGM/CJ/SA/020SGG20

PORTANT MODALITÉS DE PROTECTION DES NAVIRES DANS LES EAUX TERRITORIALES DU BÉNIN

LE MINISTRE DES INFRASTRUCTURES ET DES TRANSPORTS,

LE MINISTRE DÉLÉGUÉ AUPRÈS DU PRÉSIDENT DE LA RÉPUBLIQUE, CHARGÉ DE LA DÉFENSE NATIONALE,

LE MINISTRE DE L'INTÉRIEUR ET DE LA SÉCURITÉ PUBLIQUE,

LE MINISTRE DE L'ÉCONOMIE ET DES FINANCES,

Vu la loi n° 90-32 du 11 décembre 1990 portant Constitution de la République du Bénin, telle que modifiée par la loi n° 2019-40 du 07 novembre 2019,

vu la loi n° 2010-11 du 07 mars 2011 portant code maritime en République du Bénin,

vu la loi n° 2019-07 du 14 janvier 2019 fixant le régime des armes, munitions et autres matériels connexes en République du Bénin,

vu la décision portant proclamation, le 30 mars 2016 par la Cour constitutionnelle, des résultats définitifs de l'élection présidentielle du 20 mars 2016,

vu le décret n° 2019-396 du 05 septembre 2019 portant composition du Gouvernement,

vu le décret n° 2019-430 du 02 octobre 2019 fixant la structure type des Ministères,

vu le décret n° 2014-785 du 31 décembre 2014 portant création, organisation et attributions et fonctionnement de l'Autorité Nationale Chargée de l'Action de l'État en Mer,

vu le décret n°2016-415 du 20 juillet 2016 portant attributions, organisation et fonctionnement du Ministère de la Défense nationale,

vu le décret n° 2016-416 du 20 juillet 2016 portant attributions, organisation et fonctionnement du Ministère de l'Intérieur et de la Sécurité Publique, vu le décret n° 2016-418 du 20 juillet 2016 portant attributions, organisation et fonctionnement du Ministère des Infrastructures et des Transports,

vu le décret n° 2017-041 du 25 janvier 2017 portant attributions, organisation et fonctionnement du Ministère de l'Économie et des Finances,

vu le décret 2020-270 du 06 mai 2020 portant obligation d'une protection armée des navires de commerce en escale dans les ports du Bénin,

Considérant les nécessités de service,

ARRÊTENT

Article 1er : Obligation des navires

Il est fait obligation à tout navire à destination d'un port du Bénin d'avoir une équipe armée de protection embarquée (EAPE). A défaut, il lui est obligatoirement fourni une prestation de protection par les Forces publiques béninoises à l'entrée des eaux territoriales du Bénin. Les frais y afférents sont payés auprès du port d'escale.

Article 2 : Autorisation d'entrée avec une EAPE

Tout navire à destination d'un port du Bénin, ayant à son bord une équipe armée de protection embarquée, adresse par le biais de sa société de consignation une demande d'autorisation d'entrée dans les eaux territoriales du Bénin avec son équipe armée de protection embarquée.

Article 3 : Modalités de demande d'autorisation

La demande d'autorisation d'entrée dans les eaux territoriales du Bénin est un formulaire obligatoirement renseigné en ligne et adressé au Directeur du port d'escale 72 heures au moins avant l'arrivée du navire.

Article 4 : Interdiction de stockage d'armement au Bénin

L'autorisation d'entrée dans les eaux territoriales du Bénin, avec une équipe armée de protection embarquée, ne vaut pas autorisation de stockage d'armement au Bénin. Le navire a l'obligation de repartir avec tout l'armement de l'équipe armée de protection embarquée après son séjour au Bénin.

Article 5 : Vérification et scellement de l'armement à quai

Tout navire, autorisé à entrer dans les eaux territoriales du Bénin, est soumis à une vérification suivie d'une mise sous scellés de tout l'armement de l'équipe armée de protection embarquée.

Cette opération est effectuée à quai par une équipe des Forces armées béninoises sous la responsabilité de la Préfecture maritime.

Article 6 : Descellement à quai au départ

Au départ du navire, l'équipe des Forces armées procède, à quai, au descellement de tout l'armement de l'équipe armée de protection embarquée.

Article 7 : Escorte des navires non pourvus d'équipe armée de protection embarquée

Les navires non pourvus d'équipe armée de protection embarquée, devant entrer au port sans séjour dans la rade, sont escortés depuis la zone d'accueil jusqu'à l'embarquement des pilotes du port d'escale et leurs entrées à quai.

Article 8 : Zone d'accueil des navires

La zone d'accueil des navires est un quadrilatère délimité par les quatre points ci-après :

- Point A : 06°16'N 002° 28'E
- Point B : 06° 11'N 002° 28'E
- Point C : 06° 11'N 002° 23'E
- Point D : 06° 16'N 002° 23'E

Les mouvements des navires vers la zone sus indiquée sont coordonnés par le sémaphore de la Marine nationale.

Article 9 : Protection des navires au mouillage dans la rade

Les navires non pourvus d'équipe armée de protection embarquée et devant séjourner dans la rade, se rendent directement aux postes de mouillage attribués par le sémaphore et accueillent à leurs bords les équipes de fusiliers marins dès leurs arrivées aux postes.

Article 10 : Processus de la protection armée

La protection armée fournie par les Forces armées se présente ainsi qu'il suit :

a. dès son mouillage, l'équipe de fusiliers de la Marine nationale embarque à son bord et assure sa protection jusqu'à son accostage s'il n'a pas séjourné dans la rade et n'a pas fait l'objet d'une escorte directe ;

b. à son départ, il est escorté par un vecteur de la Marine nationale qui assure sa protection jusqu'à sa vitesse de croisière qu'il ait séjourné ou non dans la rade à son arrivée.

Article 11 : Coût des prestations

Le coût des prestations fournies par les Forces armées est à la charge de l'armateur. Il est recouvré par les services de son port d'escale. Ce coût est fixé ainsi qu'il suit :

Taille du navire (longueur)	Inspection à quai	Fourniture d'EAPE		Fourniture d'Escorte
	Mise sous scellés Levée des scellés	Forfait par navire accosté au port	Séjour en rade sur décision de l'armateur *(additionnel par jour)*	Forfait au départ du quai ou de la rade
Moins de 100 m	200 000 F CFA	350 000 F CFA	150 000 F CFA	360 000 F CFA
Plus de 100 m	200 000 F CFA	450 000 F CFA	170 000 F CFA	360 000 F CFA

3

Article 12 : Suivi-évaluation du dispositif

Un comité de suivi-évaluation est mis en place pour évaluer l'efficacité du dispositif et y apporter, si nécessaire, des améliorations. Ce comité est présidé par le Préfet maritime et est composé comme suit :

- Préfet maritime ;
- Chef d'État-major de la Marine nationale ;
- Conseiller Technique Juridique du Ministre des Infrastructures et des Transports (CTJ/MIT) ;
- Directeur des Ports (DP) ;
- Directeur Général du Port Autonome de Cotonou (DG/PAC) ;
- Directeur de la Marine Marchande (DMM) ;
- Représentant de l'Association des Consignataires et Agents Maritimes (ACAM) ;

Article 13 : Sanctions

Les infractions au présent arrêté interministériel exposent leurs auteurs aux poursuites et peines prévues par le Code maritime et le Code Pénal.

Article 14 : Entrée en vigueur

Le présent arrêté, qui abroge toutes dispositions antérieures contraires, prend effet pour compter de la date de sa signature et sera publié au Journal Officiel.

Fait à Cotonou, le 13 JUILLET 2020

Le Ministre des Infrastructures et des Transports,

Hervé HEHOMEY

Le Ministre Délégué auprès du Président de la République chargé de la Défense nationale,

Fortunet Alain NOUATIN

Le Ministre de l'Économie et des Finances,

Romuald WADAGNI

Le Ministre de l'Intérieur et de la Sécurité Publique,

Sacca LAFIA

AMPLIATIONS : PR 6 – AN 4 – CC 2 – CS 2 – CES 2 – HAAC 2 – HCJ 2 – MESGPR 2 – MJL 2 – MEF 2 – MAEP 2 – MISP 2 – MIT 2 – MDN 2 - AUTRES MINISTÈRES 17 – SGG 4 – JORTB 1.

Printed by Books on Demand GmbH, Norderstedt / Germany